BONNIE GEORDIE

BONNIE GEORDIE

THE LIFE OF TYCOON
SIR GEORGE ELLIOT

Sophie McCallum

AMBERLEY

For my father, Tony.
And to George, my great-great-great-great-grandfather.

First published 2023

Amberley Publishing
The Hill, Stroud
Gloucestershire, GL5 4EP

www.amberley-books.com

Copyright © Sophie McCallum, 2023

The right of Sophie McCallum to be identified
as the Author of this work has been asserted
in accordance with the Copyright, Designs and
Patents Act 1988.

ISBN 978 1 3981 0231 6 (hardback)
ISBN 978 1 3981 0232 3 (ebook)

British Library Cataloguing in Publication Data.
A catalogue record for this book is available
from the British Library.

1 2 3 4 5 6 7 8 9 10

Typesetting by SJmagic DESIGN SERVICES, India.
Printed in the UK.

CONTENTS

ACKNOWLEDGEMENTS

Thanks to the National Archives in Kew and especially the Weston Library in Oxford, which holds many of Disraeli's papers. I am also extremely grateful to Elizabeth Cheyne, Chairman and Archivist of Friends of Whitby Pavilion, for her knowledge of George, and for showing me the beautiful town of Whitby. Thanks also to my cousin Tim Hall-Wilson for sending me so many primary sources, and for conducting research into the stained-glass windows of Durham Cathedral. Bill Burns has given me invaluable advice and information from the illuminating website https://atlantic-cable .com/, which has provided first-hand accounts of the ship laying the first Transatlantic cable, offering a fascinating insight into the life and times of those pioneering spirits.

INTRODUCTION

In this time of massive environmental change and uncertainty, we can look back at the Victorians and observe their extraordinary ability to invent a new world. Historians have said that the Industrial Revolution was the biggest change experienced by mankind since the domestication of animals and the cultivation of plants. We have a lot to learn from the perseverance, industry and psychology of their era.

The object of this book, George Elliot, made his money in the coalfield, and when investigating his story I started off by thinking, 'Yikes! I'm an environmentalist!' But I became absorbed, not just by the social history, which is well documented in the 1844 Children's Commission into Mines, but also by the incredible enterprise of the Victorians themselves, who dragged the world into the present day.

George represents the zeitgeist of that era. He came from nothing and, as a child, used his earnings to educate himself. Driven by his pioneering spirit, he made a fortune through his vast network of mines which provided valuable Welsh steam coal to the Admiralty at a times when ruling the waves was all important. More than that, though, he also had a remarkably close bond with his men, so much so that he was known as Bonnie Geordie throughout the coalfields. He inaugurated several wire-making companies, responsible for utilising new resources found in the British Empire, and provided the first Transatlantic cable, which was laid across the floor of the Atlantic from Valentia Island, west of Ireland, to Heart's Content in Newfoundland in 1858.

Politics was another outlet for George. He became an MP for North Durham in 1868, a post he held almost continually until 1880, thereafter taking the seat in Monmouthshire from 1886 to 1892. For years he did his utmost to reform living conditions for the miners, and he was indispensable in working with Lord Aberdare on the Mines Regulation Act. He was highly valued by Prime Minster Benjamin Disraeli and was given a baronetcy in 1874 by Queen Victoria in recognition for his public services. George particularly loved being in Egypt, and after the opening of the extremely expensive Suez Canal he advised the Khedive and his finance ministers on pecuniary measures that would get the country out of bankruptcy, liaising closely with the British Government as he did so. He was also a family man, and a kind and generous friend, remaining true to his very strong roots in the county of Durham.

I

EARLY LIFE

George was born in Gateshead, probably on 18 March 1814 although some accounts list the year as 1815. He was the eldest son of Ralph Elliot, who worked as a coalminer, and Elizabeth, daughter of Henry Braithwaite, who lived in Newcastle upon Tyne. At the age of nine, while he was living at Shiney Row in Houghton-le-Spring, George began work at the nearby Whitfield Pit in Penshaw. He toiled underground as a trapper boy, opening and closing doors for the miners as they brought tubs of coal to the surface. Even at this young age, the studious George set aside a quarter of his wages for evening classes.

Reflecting on his life after his passing, George's granddaughter Florence shed some light on his beginnings:

A mere recital of the bare facts and incidents connected with the life of the late Sir George Elliot would be in itself full of interest and instruction. It is a story of progress, though a sea of difficulties in which any frail human craft would have been speedily overwhelmed, from obscurity to fame. Labour, long and arduous, was his position in early life, but from toil he won his way to high places. The pit boy became the baronet, and, in dying, has left behind him a record of good achieved for himself and his country, making his name 'on fame's eternal bead-roll worthy to by fyled'. They are the few, only, who rise, as Sir George Elliot did, from the ranks of labour to positions of affluence and of high honour. Such have all been men of great natural capacity, but they have been, also, men of untiring energy well directed, and possessing force of character which enabled them to combat successfully the sternest opposition.

The North-Country has taken its share in demonstrating the triumph of genius and labour. The history of Northumberland and Durham contains the names of several men who rose from obscurity to eminence by their own power and perseverance. Of these there has scarcely been a better type than that presented by George Elliot. He owed little to his ancestors, and, even in the days of his prosperity, he rather gloried in than sought to conceal his humble parentage. His father was Ralph Elliot, a working miner, of the 'handy' type, and therefore often employed as waste man and sinker, who migrated, even as the miners do to-day, from one colliery to another, according as advantage was to be derived from the moving. In March 1815, the Elliots lived at Gateshead, the head of the family working in the neighbourhood, and on the 18th of that month was born a son, George, who was destined to raise the family fortunes, and to bring its name to honourable distinction.

A miner's family in those days had much of the nomadic character about it, and it needed a constant struggling to bring both ends together. The house of Elliot was no exception to the rule. The little ones were many, and the means were scanty. The miner's lot then was much more toilsome and dangerous than it is even at this day. The working hours were longer, and the safeguards were fewer, and the rewards were less. Ralph Elliot was a conscientious father, so far as he had the means of providing for his flock. It is difficult now, with the magnificent educational system penetrating to the uttermost corners of our country, to realise how few and how crude were the means of learning two or three generations ago. People in the lower walks of life, if they acquired any knowledge at all of books, got it in the dame's school, of which many yet cherish somewhat mixed recollections. That was all the schooling that was given to the younger generation of the Elliots – George among the number. He enjoyed two or three years of a very rudimentary education, learning to read and write after a fashion, but little or nothing beyond. In his very early years, George went to a school at Crawcrook for a short time, while his father was engaged at the sinking of a pit in that neighbourhood, but when the family removed to Penshaw he was sent to Miss Meek's school at Shiney Row. His brief school life ceased when he had reached his ninth year. At this early age, long before even the signs of approaching manhood were visible in his frame, he donned the raiment of toil, and took his share in contributing to the family store.

His first employment was at Penshaw Colliery, where his father had obtained first the position of master wasteman and then that of overman. The colliery was the property of the Marquis of Londonderry but was in after years acquired by the lad who thus assumed the responsibility of labour with so dark and apparently hopeless a future before him. For George Elliot there was none of the frolics of youth. His work-day was fourteen hours long, and his leisure time was all too short. In later years, Sir George Elliot, harking back to the days of his boyhood, was wont to paraphrase the lines of Thomas Wilson's 'The Pitman's Pay':

> Thou knaws for weeks aw've gyen away,
> At twel o'clock o' Monday morning;
> And niver seen the leet o' day
> Until the Sabbath day's returnin.

A newspaper report from 1875 offers further insight into the making of this great man with a kind of mythical origin story:

Shiney Row occupies a sort of classic distinction among the colliery colonies of Durham, having been rendered famous by the rise of one of the most remarkable men of the time. Although the village cannot boast of having given birth to Sir George Elliot, it certainly claims him as being more closely connected with it than any other. Without doubt the hon. baronet's youthful associations all centre here. In my peregrinations through the district I heard many good stories of Sir George's early days, which proved clearly enough to my mind that the boy is father to the man. One of these must suffice as key to the rest. It shows, at least, that even in boyhood he had an *aim* and was determined to hit it. His wonderful self-reliance was always a conspicuous feature in his character, and it would appear to have strongly impressed his youthful acquaintances, and sometimes annoyingly.

On one occasion, it is told of him, that when demonstrating his superiority over the lads of the village, he chanced to give offence to a youth, known amongst the school boys under the euphonic sobriquet of 'Scrubbs', and who was by no means deficient in the organ of self-esteem. During the altercation which appears to have ensued, young Elliot so pricked his self-love that Scrubbs, provoked to retort, exclaimed, 'Aw dinnnet knaw what thou means to be at, not div aw knaw what thou means to be at.' This seems to have roused Master George's ire, for he promptly replied in the same vernacular, 'Aw'l tell ther what aw means to be at – aw never mean to stop till aw'se coal-owner of this place,' a prophecy which has since been fulfilled to the letter.

In George's day, boys and girls, sometimes aged just three or four, with many coming from the workhouses, toiled in the mines as hurriers or thrusters (collectively known as drawers), where they would work in small groups pulling and pushing the corf (a tub of coal). Hurriers were typically children of about six to eight years old, wearing a wide 'gurl' belt made from leather with a swivel chain secured on it. This belt would go around the waist of the child, with the other end attached to the cart. The child would then pull the cart from the coal face, deep underground, through mile-long passages that were often so cramped that standing was not possible even for small children. If the hurrier was fortunate, a thruster would be available. This was a younger child who pushed the cart from behind, pressing their head against the corf and often becoming bald in the process. The coal face itself was a treacherous place, inhabited by near-naked men who would load the cart with little concern for the weight of the cargo, which often exceeded 600 kilograms. Such dangerous trips were repeated for the entire fourteen-hour shift. They at least had company; women or older children often had to do the job alone.

George was a trapper boy, which was a job often reserved for those not strong enough to work with the corf. He sat in a small cutting, waiting for the sounds of the hurriers and the corf, and opened a wooden door to let them to pass. He was also instructed to open the trapdoor when it was necessary to allow ventilation to different parts of the mine. He sat in complete darkness during his fourteen-hour shift, with only the rats for company. It was not a difficult job in itself, but it could be very boring and if he fell asleep he might end up responsible for countless deaths.

In time, as mines grew larger and increased volumes of coal were extracted, ponies driven by coal drivers would take the coal to the surface. These coal drivers would have been about George's age – boys of about ten to fourteen years – but this was a long way off. At the time an observer remarked,

The transport of coal is very hard labour, the stuff being shoved in large tubs, without wheels, over the uneven floor of the mine; often over moist clay, or through water, and frequently up steep inclines and through paths so low-roofed that the workers are forced to creep on hands and knees.

The strongest teenagers or men were employed as getters, tasked with cutting the coal away from the seam using a pickaxe. They were the only workers underground who were permitted candles or safety lamps in order to see what they were doing; the others worked in total darkness. When a seam of coal was found it was extracted horizontally, so miners were called upon to lie down in very tight conditions while they toiled.

There were several serious dangers in the mines. Flooding was an ever-present risk and children would frequently work in water that reached up to their thighs. This was greatly remedied by the innovation of Thomas Savery in 1698, which was further developed and improved by Thomas Newcomen in 1712 and later by James Watt in 1766, whereby steam pressure was deployed to force water up and out of the mine. Financial constraints imposed by miserly coal owners prevented such machinery from being deployed en masse, however.

Further risk came from the poisonous and explosive gases that lurked everywhere. Gases in coal mines were called 'damps', which comes from the German word for 'vapour'. Firedamp was a collective name for the flammable gases, such as methane, found in bituminous coal. With naked flames from candles everywhere in the mines, explosions were a much-feared occupational hazard. Other gases present were chokedamp or blackdamp, which was a lethal combination of carbon dioxide, water vapour and nitrogen. Whitedamp was composed of carbon monoxide and other combustible gases. Stinkdamp, or hydrogen sulphide, was both poisonous and explosive, and smelt of rotten eggs. Afterdamp, a mixture of carbon monoxide and other gases, occurred after an explosion of firedamp.

Although the dangers of firedamp in the coalfield had been investigated by the Royal Society from 1677, it was not until 1815 that Sir Humphrey Davy invented the safety lamp, which supplanted the candle. This lamp, known as the 'Miner's Friend', had a wire gauze that prevented the lamp's heat from escaping and igniting any gases present in the atmosphere, but it did not give off a great deal of light and the miners consequently struggled with visibility.

Miners famously found a solution to the problem of dangerous gases by taking a canary to work. If the canary died, then dangerous gases were present and it was time to leave the pit. Another safety practice was the cutting of two shafts to link tunnels. A fire was lit at the bottom of one shaft, drawing fresh air down the other. In 1807 John Buddle invented an air pump which increased airflow, but it was costly and therefore not widely used.

The risk of firedamp increased with depth. Firedamp would gather in the rock face, exploding with great force when the coal was struck. Concentrations of firedamp between 4 and 16 per cent were flammable, and even the use of Davy lamps did not help, as mere sparks from metal tools striking the rock would ignite the gas. Firedamp was an ever-present danger, killing hundreds of miners at once, exploding with tremendous force and blasting men up shafts 200 yards deep as if from a cannon, shaking the solid structure of the earth around the pit like an earthquake. Even with adequate ventilation, firedamp explosions would occur when the air doors were left open and through the flouting of other safety precautions. Time and time again, inquests into such incidents ruled against the miners and in favour of their employers. If a miner survived an explosion, they were generally left with horrible burns that scarred their body.

Lighting was a constant problem underground. Miners devised some unusual techniques to shed light, including rotten fish, which emitted a phosphorescent luminescence. By 1842, improvements made to the Davy lamp by the Chamber and Werneth Company in Oldham allowed it to give a more brilliant light than before. Electric lamps, which were much safer and more reliable, only came into use in the 1890s.

Chokedamp was not as fatal as firedamp, and could be detected by the dwindling of a candle flame. It was an asphyxiant, lowering the oxygen in the atmosphere and eventually resulting in death. This toxic blend of gases was released from the coalface as it absorbed oxygen from the air. It was usually found in badly ventilated mines, or those which had been out of use, and was a particular hazard in the spring and summer as coal emits more carbon dioxide in hotter temperatures. Other factors included the level of exposed coal and the variety of coal.

The symptoms of chokedamp were slow to materialise: dizziness, tiredness, lack of coordination and feeling faint. These crept up and could easily be mistaken for fatigue. Once the symptoms accumulated, however, asphyxiation and unconsciousness could follow within moments.

Another common accident resulting in death happened when the getter stopped working with his pickaxe. His movements dispelled sulphur gases, and when he stopped they gathered thickly and were ignited by his candle. Both getter and drawer were frequently killed in such instances. Sulphur was often found on the Tuesday following Monday payday, which the miners had off, as it the gas had a chance to accumulate in the mines. It acted on the body by swelling it, and the miner would doze off as if he were asleep.

The trapper was therefore all important in removing the threat of gases by providing a basic air conditioning system. In opening the doors they created a draft which could diffuse clouds of dangerous fumes. It was said that the owners of the mines also liked this system as the closed doors would isolate an explosion, sparing the rest of the mine and protecting their investments. However, the system was too crude to be effective and many deaths were caused by a lack of proper ventilation. Some coal owners would sooner make the workers toil in dangerous conditions than pay for proper ventilation. Only the most experienced colliers were brave enough to enter such pits. The younger miners did not dare risk it as they had no experience of dealing with the gases.

Another major danger to miners was the collapse of a tunnel. The structure of the mine was only held up by wooden beams, and with the enormous weight of the ground above it a mined coal seam often gave way. A parliamentary report on deaths in coal mines described the many ways a miner might lose his life – being crushed in a collapse, falling down a mine shaft, suffocating from poisonous gases, being hit by tumbling coal, falling out of a 'bucket' bringing him up after a shift, explosions, and finally being run down by a cart.

A first-hand account from Arthur Eaglestone, a miner at the New Stubbin colliery in Rotherham, recalls one such tragedy:

He was quite dead when we found him – squashed by a fall of stone. The smell of blood was everywhere, a slaughterhouse reek, and sickening. His body seemed to be almost completely covered by the tremendous boulder that had fallen. Only one hand was visible and the upper portion of his forehead – nothing else. His mate wept openly, speaking sometimes with a strangled incoherence, a clucking babble of words, but no one took much notice, for the man was dead enough, and the only help we wanted was in rolling the stone away. If trembling went for anything, we were all goosey. I didn't want to stay. I didn't want to look on IT! My heart sickened at the thought of all that mangled flesh. So, craven of spirit I slid my lamp around my thigh, seeking comfort in the shadow that my body threw upon the rock.

'If the jack's not coming,' said Morgan, 'we'll try again. All together … now then … heave ho!' We gathered in, crowding against the protuberance like rugby players in a football scrum. Our fingers touched the hard rock gingerly at first, as though it held some sacred quality, but Morgan's harsh controlling voice with 'Now then, no playing! put to it … put to it! … all ready? … heave!' inspired us to lofty physical effort … and so we pushed … the stone lifted. 'Shove the block in,' hissed Morgan, 'quick! Now … heave! ah!'

Oh! He was there all right. The first thing that I saw was the sloppy pool of dirt that was his body. And then his face all coal and wax in the midst of which two eyes wide open, staring, shone strangely golden in the swinging lamplight, with the same illumination that a cat's have, in firelight, or sunlight.

And then again in thick and cloying waves, the stench of blood … the indrawn sighing of the rescuers … the thin insistent hissing of the face itself… S-S-S-S-S.

The roof of a mine could collapse at any moment, with any stroke of the pickaxe. Miners became fine-tuned to the sound of their axe on the roof, especially when removing posts from works in order to get at more coal. Again, miners were blamed for their folly in the consequence of an accident, but it was shown that where employers provided more wood there were fewer collapses.

Of all deaths in the mines, most took place when miners were travelling up and down the shaft. The coal owners were responsible for maintaining and replacing the rope on which men were lowered down to their work, and the rope often had to get into a very poor condition, with broken strands, before a new one was bought. The owners of deeper mines tended to replace their ropes more often, due to the considerable expense of work disrupted by a broken one, whilst in the smaller mines, with less of a drop to the coal face, owners tended to be more relaxed.

In some mines, the sides of the shaft were bricked the whole way down and set with mortar. This prevented large rocks from coming loose and falling down the shaft onto the workers, which was a common cause of death. This was reported to increase costs by only a small amount, even in shafts 1,300 feet deep, and effectively eliminated such accidents. Corfs being lifted through the shaft were another danger, and great care was taken not to overload them. When the coal reached the surface, implements such as chains and sliding stages prevented it from tumbling back down the shaft. Corfs descending and ascending the shafts frequently collided, however, sending coal tumbling down to the pit-eye beneath and causing death or injury.

There were very few families in coal-mining communities who had not suffered a death from accidents in the mines. Some families were bereaved of three or four family members. Accidents occurred every few weeks at any given pit.

In fact, the frequency of accidents meant they were seldom taken seriously. One boy recalls:

I have been hurt, but colliers don't make any account of being hurt, unless their bones are broken. My brother had his leg hurt by the roof falling, and his leg has been taken off; he is getting well now very nicely.

Another lad says:

> It was very wet in the pit, it always came up to my ankles, and I had to wade up to my knee in water and sludge when I went in and came out of the pit. I had my head hurt badly; it kept me in bed ten days; I lost the sight of one of my eyes by some red water falling into it; cannot see with it now at all.

Many of these accidents could have been prevented by more care from the coal owners:

> I remember a case two years ago where a boy of ten years of age was managing an engine whilst five men were coming up, and he let the engine wind them over the head gear and they were all killed. I was foreman of the coroner's jury at the time, and we represented the danger of entrusting engines to the care of such young children to one of the masters who attended the inquest, but he said the engines were just as safe in the hands of children as grown-ups ... It is still common to employ children as engineers.

In the findings of the Children's Employment Commission (Mines) 1842, one source wrote:

> I was myself unfortunate enough to witness a fatal accident, when a little boy, of 10 years of age, was killed in this manner. The chair which held the tub in which he was standing came in contact with the descending chair; the concussion threw the child out of the tub, and it fell 60 yards down the pit, and was literally dashed to pieces.

The aforementioned 1842 commission was an official investigation into the conditions of those employed in the coal mines of Great Britain and Ireland during the years 1840 to 1843, especially children. The findings were printed in the *Parliamentary Papers* and was made up of 4,108 testimonies covering 15 volumes. The commission was led by Lord Anthony Ashley-Cooper, 7th Earl of Shaftesbury, and was requested by Queen Victoria after a devastating accident in 1838 at the Huskar Colliery in Silkstone near Barnsley, where twenty-six children were killed: twelve girls aged eight to sixteen, and fourteen boys aged nine to twelve. The tragedy came about when a terrific storm swelled a stream so that it flowed into the mine's ventilation system. Shaftesbury, who was particularly passionate about children's safety, had pushed hard for the royal commission.

In 1841, there were 216,000 people working in Britain's mines. The commissioners travelled throughout the coalmining regions, often investigating in the face of hostility from the coal owners. The Children's Employment Commission's final report was published in May 1842, complete with engravings illustrating the conditions workers had to endure. Witnesses were produced and thoroughly cross-examined.

Betty Harris, aged thirty-seven and listed as a drawer, said,

I have a belt round my waist, and a chain passing between my legs, and I go on my hands and feet. The road is very steep, and we have to hold on by a rope; and where there is no rope, by anything we can catch hold of ... I am not as strong as I was and cannot stand the work as well as I used to. I have drawn till I have had the skin off me; the belt and chain is worse when we are in the family way [pregnant].

Patience Kershaw, aged seventeen, gave her own account:

The bald place upon my head is made by thrusting the corves ... I hurry the corves a mile or more underground and back; they weigh 3 cwt [150kg] ... The getters that I work for are naked except for their caps ... Sometimes they beat me if I am not quick enough.'

A sub-commissioner inspecting a young girl at a pit near Huddersfield said, 'I could not have believed that I should have found human nature so degraded.' Mr Brook, a surgeon living close to mines, acknowledged that they could not have believed that such a system of un-Christian cruelty could have existed. Talking about one young girl, he said, 'She stood shivering before me from cold. The rug that hung about her waist was a black as coal, and saturated with water, the drippings from the roof.'

The commission soon realised that the majority of children started work in the mines at the tender age of six or seven, sometimes even younger. However, where coal seams were thicker, larger children aged from fourteen to eighteen were required as drawers because the tubs of coal were too heavy for young children. Coal owners had discovered that it was cheaper to expand the main roads in the mine, using tubs that held 4 to 6 cwt (200–300 kg) of coal. Young children were of course still required for very thin seams of coal as they were small enough to move through the passages. Coal seams could be at any gradient, some forming an almost horizontal angle, while others rose 2 foot in every yard. This impacted heavily on the children's work.

Tubs weighed around 4 cwt (200 kg) on the inclines. On the horizontal, drawers took two tubs, weighing 8 cwt (400 kg) in total, at a time. With a clay soil, the mud was quarter of a yard in depth, and often the water reached the knees of the drawer. Drawers wore clogs when the mines were full of water and mud but took them off to climb the inclines as they needed to be as light as possible for this work.

Miners reported, when asked, that drawing was much harder than getting, and more often than not the efforts so stunted drawers that they did not grow to their full adult strength. The poor air quality made the work even more arduous, especially in the cramped conditions.

Drawer Benjamin Barry tells the commissioner that he pulled the corf 800 yards on rails horizontally back and forth, and then 360 yards on an angle, both up and down, with no rails. He did this twenty times a day, amounting to a daily total of 30,400 yards or more than 17 miles. And there were consequences if he slowed, as one collier reported:

> If a drawer is too weak, he gets beat: the weakest are always worst off, and get worse beaten; they have not the strength to do the work in time, and the colliers thrash them.

The mines were usually accessed by the shafts, but they could also be entered via galleries which were bored into the slope of a hill. If the mine was less than 350 feet below ground level, two shafts were generally sunk. Over time, as more tunnels were bored on a grid system, the mine took on the look of a subterranean town with a multitude of roads and lanes.

Commissioner John L. Kennedy reports on such an arrangement:

> At one pit I visited there was a furnace in the bottom of the engine-shaft, and we descended through the smoke, as it were, down the chimney. On approaching the bottom, figures became visible, flitting to and fro about the furnace. The darkness around us, and the deep red flames seen through the smoke, together with the shrieks and voices of people below, gave a forcible idea of a descent into Pandemonium.

Trappers were still used extensively in mines in 1842, with tunnel networks stretching for 60 to 70 miles. A door on a spring, opening both ways, was proposed, but miners were wary about this unmanned portal breaking down, adding to the danger of firedamp.

A commissioner reported on the life of a trapper:

> This occupation is one of the most pitiable in a coal-pit, from its extreme monotony. Exertion there is none, nor labour,

further than is requisite to open and shut a door. As these little fellows are always the youngest in the pits, I have generally found them very shy, and they have never anything to say for themselves. Their whole time is spent in sitting in the dark for twelve hours and opening and shutting a door to allow the waggoners to pass. Were it not for the passing and repassing of the waggons it would be equal to solitary confinement of the worst order.

He went on:

In endeavouring to describe the various employments of the children and young persons engaged in coal mines, I found reason to believe that no *words* I could use would convey to others, impressions, similar to those, which ocular inspection had given to myself.

The working conditions of the collier were related to the thickness of the seam, the water level in the mine, ventilation and, finally, the working practices of the collier. Coal getters worked in spaces as small as 18 to 20 inches in height, with the average seam height being around 3 feet. A common position was to rest one's chest on one's thigh, with the head touching the knee. In seams that were 14 inches thick, miners were forced to lie on their sides whilst they used their pickaxes. They had 'pilchers' and 'arm-patches' to stop their skin from being scraped off. As a result of such friction, the skin of their arms became as thick as it was on the soles of their feet. In the deeper levels, miners worked in extremely wet conditions, with water filling up half the tunnel. Even with the benefit of ventilation, most miners suffered from asthma in old age.

The plight of women was singled out by one commissioner, who wrote that 'no argument is necessary to convince a person who has witnessed females at work in the mines, that it is an employment very ill-suited to their sex. Betty Wardle, a housewife from Outwood, near Lever, was asked by the commissioner,

'Did you work in the pits whilst you were in the family way?' She replied, 'Ay, to be sure. I had a child born in the pits, and I bought it up the pit-shaft in my skirt.' Stillbirth in children was fairly common in the pits, and many children died soon after they were born.

Betty Harris, the thirty-seven-year-old drawer we encountered earlier, has more to say:

> I have been hurt once: I got on a waggon of coals in the pit to get out of the way of the next waggon, and the waggon I was on went off before I could get off, and crushed my bones about the hips between the roof and the coals: I was ill twenty-three weeks.

Girls were beaten as much as the boys if they did not behave themselves. Some drawers were thrashed so badly with a pick-arm, belt or foot that they had to give up drawing. They were often so tired at night that they would fall asleep before having their supper. Frequently, they would lie down at the roadside to snooze before walking the 2 or 3 miles home, their parents finding them cuddled up on the verge. Save for Sundays, these children did not see daylight the whole winter through.

Drawers took turns to deliver the coal, and if they missed their turn to fasten their corf's hook at the pit-eye, it resulted in lower wages for both the drawer and the collier that day. It was commonplace for single men to have relationships with girls that drew for them, sometimes living 'in tally' –as an unmarried couple. John Oldham, aged fifty-eight, said in 1841, 'There are a great deal of bastards; I have frequently known the women to be in the family way with the colliers they draw for and others.' He continues:

> 'Yes, [drawing is hard work] for young children and women, especially when the women are in the family way. I have known women draw coal in a pit to within a few days of their time; ay, I have known them work the very

day they were brought to bed. My wife worked at drawing till the day before she was brought to bed for her last child. There are women now working in the pit who are in the family way.

The inspection of the mines itself was a fairly torrid affair. John L. Kennedy recounts:

I doubt whether Inspectors could be found who would faithfully descend shafts two or three times the depth of the height of St. Paul's, and amidst wet and damp and noxious gases crawl or allow themselves to be waggoned through miles of dark drains and subterranean caverns, with the chance of the roof falling on them or being burnt by explosion, *to see that all was right*, and not act on the easy assumption that it was so.

Coal owners themselves did not stray underground, allowing their underlookers or foremen, many of whom were former colliers, to do that job for them. Accidents, by their very nature, destroyed most of the evidence of who was to blame. In a firedamp explosion, for example, the doors would be annihilated, the wagons blown to smithereens and the roof would cave in due to the air void in the mine. Nothing was left, and it was all too easy to blame the miners. Correspondingly, if a boy was hurled from a tub ascending the shaft by a collision with a tub going down, evidence showing that a crash had taken place had to be supplied before his brutal death could be found to be anything other than 'accidental death' by the coroners. Miners were very scared of coming forward to speak out against the coal owners, and time and time again their negligence was blamed rather than pit owners taking responsibility for safety.

At the time of the commission in 1842, children and young people worked on average an eleven- or twelve-hour day, starting at five or six in the morning and finishing by five or six in the afternoon. However, the Monday pay day was considered a holiday by the men, who often stayed away from the pits until midweek. Coal owners were exasperated by this, with boatmen and banksmen waiting around for the men to arrive. Both getters and their drawers worked doubly hard when they did return to work in order to make up lost time, and the drawers quickly became exhausted. Coal owners devised a plan whereby the colliers were only allowed to send up a fixed number of tubs per day, meaning the banksmen and boatmen were not standing idle and the drawers weren't so overworked.

However, during winter, when the market for coal was at its peak, the length of the working day was increased to thirteen hours and it was full steam ahead. Some pits were busy throughout the night to meet demand and to maximise profit. Young people and children were expected to work these shifts too. Even so, it was found that the night shift produced about a fifth less coal than the day shift. Leases for the mines were taken out on the premise that night work would go ahead, and therefore the expected profit included this night shift. A few miners reported seeing ghosts in the mines. This was a time when even the roughest men, who kept fighting dogs and cocks, believed in the supernatural. Unsurprisingly, more ghost sightings were reported at night.

The colliers were said to eat well. For breakfast they had bread and butter and coffee, the latter often without milk as this was a very expensive commodity. For lunch, which was eaten at work, the collier had bread and cheese, taking a bite whenever he had the time. This lasted until six in the evening, when he tucked into meat and potatoes. A great favourite was bacon, and sometimes the collier got beef and mutton. Although there were a few minutes here and there when the muscles could rest, the great majority of the day was spent in physical exertion. In the well-run mines, where half an hour to an hour was given for lunch, the cases of fatigue in those aged ten or older was greatly reduced.

Miners, including women and children, washed their faces, neck and hands after work, but seldom washed their bodies, instead allowing the dirt to be rubbed off by their clothes. They suffered terrible boils as a result and children were prone to inflamed eyes, leading to some surgeons promoting the idea of a bath next to every pit, which would allow the miners to wash after work.

Miners received quite generous holidays, including a fortnight's break over Christmas, a week for Whitsuntide and the odd week here and there for local summer festivities and the races. With time off due to accidents, colliers, including the children, worked fewer hours than those employed other industries. Although the miners were paid every month, there was a small payment every fortnight.

Boys were paid according to their age. When a boy was below ten years of age, he was considered to be one-eighth the strength of a man and was paid accordingly. By the age of ten, he was considered two-eighths and his pay was increased. At thirteen he rose to three-eighths. Two years later, he reached one-half strength. At eighteen, he was considered three-quarters a man and could become a getter. A girl was considered to be one-half a man's strength at the age of sixteen, but females would never work as a getter, only being suited the role of drawer. Drawers said that they had enough to eat, but they were often dressed in rags and did not have a change of clothes.

A bond was a common agreement between miners and the agents of coal owners. For example, a miner would be hired for eleven months, at which point he could borrow £3 to £4 for himself and a further £2 or £3 for his son if he was capable of using a pick. The owner took 3 shillings a week from his pay and 2 shillings from that of his son. If the son did not pay his debt the father had to make it up, essentially pawning his son's labour. These kinds of agreements were only made with the getters, never directly with the drawers. The agent could use the bond to his advantage by making men work in wet places, which were the least popular areas and which in other circumstances saw the miners paid at a higher rate. It was not a good deal for the miner, who often spent the £3 or £4 in the alehouse with the debt hanging over him until he paid it back. The alternative was debtor's prison.

Managers found the colliers worked better as they were paid by the quantity of coal that they extracted, in contrast to the pony drivers and the drawers who were paid a daily rate and were therefore harder to manage. Those who were paid by the day, both girls and boys, were beaten more often for slacking, whereas the colliers were more motivated given their quotas.

A commissioner reports on one case of beating;

> Mr. Milner examined this boy and found on his body twenty-four to twenty-six wounds. His posteriors and loins were beaten to a jelly; his head, which was almost cleared of hair on the scalp, had the marks of many old wounds upon it which had healed up; one of the bones in one arm was broken below the elbow, and, from the appearance, seemed to have been so for some time.
>
> The boy, on being bought before the magistrates, was unable to sit or stand, and was placed on the floor of the office, laid on his side on a cradle bed … The boy's arm had been broken by a blow with an iron rail, and the fracture had never been set. [Brierley] had been in the habit of beating the boy with a piece of wood, in which a nail was driven and projected about half an inch … The brutal master had kept him at work as a wagoner until he was no longer of any use, and then sent him home in a cart to his mother, who was a poor widow, residing in Church Lane, Rochdale.

Children from the poorhouse were bound to the mining trade more than any other business. They suffered horrific abuse, sometimes nearly fatal, as there was no one to look out for them. In one case, John Halliwell, the overseer of Oldham Colliery, described how three pauper children, aged ten to twelve, had not brought lunch and were accused of stealing food from the other children. One of the strongest men took the boys one by one and held their heads between his knees, telling each of the eighteen boys in the pit to thrash them twelve times with a 'cut'. One of the boys had been beaten so severely that a doctor said he would not survive. The other boys in the pit had been forced to beat them lest they receive the same punishment. The older men in the pit were philosophical, as this practice had been going on since they were children and they thought it was perfectly sound and reasonable. The person overseeing the mines would rarely get involved to stop cases like these, saying that they did not think they had the right to do so. The coal owners showed little interest in the welfare of children and young people once they had done their day's work. They did not provide any entertainment or recreational facilities for them.

Mining families were renowned for their lack of education. Very few ever went to church. The pubs were full on Saturday nights and Sunday mornings, and fighting and other disturbances were a common occurrence. Unsurprisingly, moral virtues were hard to find in the mining community. Some of the women kept the same pace as the men in the pubs, whilst others were a little less disorderly. They were generally nonchalant about having illegitimate children, many of whom were taken on by the workhouse. Some girls were described as 'glorying' in their illicit child and showing no sense of decency. Girls got better wages in the mines than they did in weaving cottages or factories, which some attributed to the high rate of unmarried pregnancies. It was unheard of for a girl who had started down at the pits to be re-employed as a servant because she had learnt such coarse ways.

Charlotte Chiles worked at the Graig colliery in Merthyr Tydfil, Wales, where she drew, landed and weighed coal, earning 40 shillings a month in the 1840s. She had previously been employed in kitchen work by Lord Kensington near Carmarthen and would earn 60 to 70 shillings a year with lodgings. She said,

> I prefer this work as it is not so confining, and I get more money ... I cannot save money now: but I get more dress and more liberty. I work 12 hours daily ... The work, though very hard, I care nothing for as I have good health and strength.

Even though the colliers and drawers were taking home quite a lot of money each week – for example, a collier around 25s to 30s and his wife, a drawer, 14s to 15s in the mid-nineteenth century – their homes were poverty-stricken. Children would be infested with lice.

By and large the colliers of the 1840s were illiterate, having started their life in the mines at too young an age to gain an education. Coal owners preferred somewhat educated men, however, saying that they needed a lot less supervision. Uneducated colliers were strong with the pick but were said to make trouble and did not understand what was required of them. In a survey of 1,113 males between thirteen and eighteen years old, only 24 per cent could write their names. With females in that age range, just 1.4 per cent could do it. With boys under thirteen years, only 8.5 per cent could write their names, and females in that age range could write their name in 2.2 per cent of cases; even then, this was significantly lower than in the general population.

Colliers were described as deceptive cheats and vagabonds, drinking whenever they had the money and enjoying fighting and wrestling on a regular basis. Other working-class people, such as handloom weavers, living in the same communities as the colliers, refused to mix with them, even though colliers enjoyed higher wages. The colliers were regarded as savages. According to reports of the time, although colliers worked fewer hours, they spent their free time getting drunk and engaging in sexual acts in a brutal and heartless manner. They amused themselves with dogfighting, cockfighting, foot races and gambling. Handloom weavers on the other hand, earning less than colliers at 7 to 8 shillings a week, were renowned for their love of learning, often owning books and scientific apparatus. It was thought that improvements in the moral virtues of colliers could only be established by legislation encouraging the education of their young.

As for their living quarters, a cottage and a fire was a considerable expense for the mining family, costing 5 shillings a week throughout the whole year. They often had houses containing only one room on the ground floor and one room above, frequently sleeping with six or seven in the same bed. Ventilation was minimal as houses were usually back-to-back, allowing no through draft, with two or more families sharing the same house. There was no sewage system and the streets became filthy. Harmful odours caused by animal and vegetable waste, mixed with human waste, gathered and putrefied in the street, causing severe fevers. Privies did exist, but not enough to fulfil demand and in such a squalid condition that people did not use them. While a farming village contained clean, orderly houses with well-tended gardens compared with the filth and sewage found outside a collier's door, often with a ferocious bulldog snarling inside.

In conclusion, commissioners for the 1842 report found that children were going to work down the mines at too young an age, causing stunted growth and depriving them of an education, creating a class of men who were brutal, uneducated and dangerous to the state. The report mentions the difficulties in inspecting the mines, when the children were taken down before daybreak and stayed in the mines until after dark. Such was the underground labyrinth; it would be very easy for children or adults to escape detection. The only way children could be prevented from going down the mines was to make education compulsory until ten years of age. They would then have the opportunity to meet children from different backgrounds and learn from their examples. Commissioner John L. Kennedy in the Lancaster coalfield estimated the cost of a schoolmaster to be £1 a week. Some 1,100 children were educated by Norwood School, where the cost of the teachers was £700 per year, which worked out as 12 shillings per child every year. More teachers would create a bigger knowledge base.

It was decided that paying the miners on a fortnightly or monthly basis was not a good idea. Miners bought goods at the shop on credit if they had spent all their money, and the shopkeeper would raise the price of the goods to safeguard himself. If he were to be paid in cash, the prices in the shops would be lower. The parish would not allow mothers of coalmining families any money, saying that they got better money than weavers and could look after the child themselves. Both mothers and fathers could cause their children to starve by spending their wages in the alehouse. On Sundays, rather than going to church, they would walk in the fields, sometimes poaching and catching birds, enjoying the sunshine whilst they could.

Housing was another area that the commissioners felt required improvement:

The houses are usually filthy; there is no attention to whitewashing or ventilation; paper and rags are often pasted and stuffed into broken windows; the beds and bedding are generally poor; they are in the habit of pawning their clothes, and in some instances I have known them to pawn the bedding, bed-clothes, and even the coats of their backs, when they have been on the 'spree'.

I have seen cases where the father of the family has been in receipt of tolerably good earnings, but from dissolute habits the children are neglected, and almost in a state of starvation. In some of the houses the man, wife and five children, sleep in one room, and often after they are grown to be men and women.

The colliers have very few comforts; their houses are generally wretched and are most always without furniture. They are constantly changing their houses, and there are always repairs required in houses that colliers have left than any other tenants – in fact they use their houses as if they were a coal pit.

... On entering [one man's] house, we found him looking out of his window as dirty as it was possible for a man to be. In the room there was no furniture, with the exception of a chair and a rude bench to sit down on. There was a bundle of sticks lying on the floor; on looking into the cupboard there was not the least vestige of anything edible. Mr. Ainscough, the assistant overseer, who accompanied me, stated that he had been informed by the employer of his two daughters that they received 16s per week, and the man himself admitted that he was in the receipt of £1 10s per week. The house was miserable and wretched in the extreme.

On visiting another family, commissioner John L. Kennedy reported that 'a neighbour told me that she had given the father some bread and cheese the other day, and that he ate it like a tiger. When I gave the child something to buy food, I was quite shocked with the look of ravenous joy with which they received it.' However, there were a few exceptions:

> For a married couple and their five children – with three children working, and an apprentice. The house is neat and well-kept with three clean beds. This is a rare, other extreme of a collier's house. The joint earnings of the family come to £3 a week, which is spent on the following each week:
>
> Rent, 2s, taxes 3d, fire 1s 9d, 5lb sugar 2s 11d, 6oz tea, 1s 8d, 46oz flour, 9s 7d, 1oz meal, 1s 8d, 4oz butter, 4s 2d, 2oz cheese, 1s 4d, 6oz potatoes, 2s, 9oz meat 4s 8d, 7 quarts milk, 1s 2d, 1.5oz soap 7 ½ d, pepper and mustard 1d, thread and worsted, 6d, blacking and black lead 1 ½ d, Bristol brick 3d, Sick Club, 2d.
>
> This amounts to £1 16s 9d each week.
>
> This means that each member of the family would receive almost 2lbs 8oz of solid food daily and they are able to keep a pig.

The 1842 report understandably shocked the country. In one coal mine, out of the total of 349 deaths in one year, fifty-eight were children under the age of thirteen.

Friedrich Engels sums up the situation in his influential book *The Condition of the Working Classes in England* (1845):

> In the whole British Empire there is no other occupation in which a man may meet his end in so many diverse ways as in this one...
>
> The coal-mine is the scene of a multitude of the most terrifying calamities, and these come directly from the selfishness of the bourgeoise. The hydrocarbon gas which develops so freely in these mines, form, when combined with

atmospheric air, an explosive which takes fire upon coming into contact with a flame and kills every one within its reach. Such explosions take place, in one mine or another, nearly every day; on September 28th, 1844, one killed 96 men in Haswell Colliery, Durham. The carbonic acid gas, which also develops in great quantities, accumulates in the deeper parts of the mine, frequently reaching the height of a man, and suffocates everyone who gets into it. The doors which separate the sections of the mines are meant to prevent the propagation of explosions and the movement of the gases; but since they are entrusted to small children, who often fall asleep or neglect them, this means of prevention is illusory. A proper ventilation of the mines by means of fresh air-shafts could almost entirely remove the injurious effects of both these gases. But for this purpose, the bourgeoisie has no money to spare, preferring to command the working-men to use the Davy lamp, which is wholly useless because of its dull light, and is, therefore, usually replaced by a candle. If an explosion occurs, the recklessness of the miner is blamed, though the bourgeois might had made the explosion well-nigh impossible by supplying good ventilation. Further every few days the roof of a working falls in and buries or mangles the workers employed in it. It is the interest of the bourgeois to have the seams worked out as completely as possible, and hence the accidents of this sort. Then, too, the ropes by which the men descend into the mines are often rotten, and break, so that the unfortunates fall, and are crushed...

He goes on:

In nearly all mining districts the people composing the coroner's juries are, in almost all cases, dependant on the mine owners, and where this is not the case, immemorial custom ensures that the verdict shall be: 'Accidental Death'. Besides, the jury takes very little interest in the state of the mine ... Hence, and from the tender age at which children are put to work, it follows that their mental education is wholly neglected. The day schools are not within their reach, the evening and Sunday schools' mere shams, the teachers worthless. Hence, few can read and still fewer write ... To church they go seldom or never; the clergy complain of their irreligion as beyond comparison. The categories of religion are known to them only from the terms of their oaths. That the overwork of all miners must engender drunkenness is self-evident. As to their sexual relations, men, women, and children work in the mines, in many cases, wholly naked, and in most cases, nearly so, by reason of the prevailing heat and the consequences in the dark, lonely mines may be imagined. The number of illegitimate children is here disproportionately large and indicates what goes on among the half-savage population below ground; but proves too, that the illegitimate intercourse of the sexes has not here, as in the great cities, sunk to the level of prostitution. The labour of women entails the same consequences as in the factories, dissolves the family, and makes the mother totally incapable of household work.

Victorian prudery found these accounts of near-nakedness, with girls and women wearing trousers and working topless in the company of the men and boys, more offensive than their actual working conditions. On 7 June 1842, the Earl of Shaftesbury made a speech in Parliament requesting leave to bring in a bill standardising the employment of women and children in mines, based mainly on the evidence presented by the commissioners:

Sir, it would be impossible to enlarge upon all these points; the evidence is most abundant, and the selection very difficult. I will, however, observe that nothing can be more graphic, nothing more touching than the evidence of many of these poor girls themselves. Insulted, oppressed and even corrupted, they exhibit, not unfrequently, a simplicity and a kindness that render tenfold more heart-rending the folly and cruelty of that system that has forced away these young persons, destined, in God's providence, to holier and happier duties, to occupations so unsuited, so harsh, and so degrading...

Surely it is evident that to remove, or even to mitigate, these sad evils will require the vigorous and immediate interposition of the legislature. The interposition is demanded by public reason, by public virtue, by the public honour, by the public character, and, I rejoice to add, by the public sympathy: for never, I believe, since the disclosure of the horrors of the African slave-trade, has there existed so universal a feeling on any one subject in this country, as that which now pervades the length and breadth of the land in abhorrence and disgust of this monstrous oppression. It is demanded, moreover, I am happy to say, by many well-intentioned and honest proprietors – men who are anxious to see those ameliorations introduced which, owing to long established prejudices, they have themselves been unable to affect. From letters and private communications which I have received on the subject, I know that they will hail with the greatest joy such a bill as I shall presently ask leave to introduce.

The Mine Act 1842 was subsequently passed and ensured, among other things, that no female nor boy under ten was to be employed underground but that parish apprentices between the ages of ten and eighteen could continue to work in the mines. There were no conditions set on hours of work, and enforcement of the Act could only be made on the premise of monitoring the 'condition of the workers'. Women were generally unhappy about the passage of the Act, which blocked off a much-needed source of income. Kennedy noted that 'one old miner in particular, with a large family, all girls, told if his three daughters, which he employed with him in the mine, were drawn out, the whole family must become inmates of the Union workhouse, as his own labour and exertions could not support them'. It was such a problem that for several years the Act was ignored.

To return to our protagonist, George Elliot was twenty-seven years old at the time of the Children's Employment Commission in 1842. However, he had been involved in strikes over the length of the working day as early as 1831, becoming a union leader when he was sixteen years old. Workers in the coalmining industry were pushing for new regulations to enforce better working conditions, likening their case to the abolition of the Atlantic slave trade and alluding to the French Revolution. Similarities were drawn between the pitmen and the slaves by William Scott, who reached out to the sensibilities of the British Public in 1831:

> After the poor but brave Pitman descends to the bottom of this horrific cavern, he has however to walk, or rather crawl a mile, or perhaps two or three, through subterraneous vaults, where the 'pestilence that walketh in darkness' hangs thick and heavy around him, and at every step he is liable to be crushed to pieces by some ponderous body from the roof; whilst the slightest inadvertency or casualty in the most distant part of the Pit, might, and often has in an instant involved him and his fellow slaves in destruction, and in many instances left the mangled fragments of his body undistinguishable from those of scores of his fellow sufferers.

The coal owners replied to William Scott's appeal, saying that they were able to find new recruits who were happy to work the hours and quickly gain the skills that the striking workers and seasoned coal getters already had. They mocked the powerless colliers, saying that they already had the right to work wherever they liked. The pitmen replied by appealing to the public, stating, 'It certainly does not become gentlemen to act with such dominion over their fellow-creatures, who are the instruments of accumulating them princely fortunes, at the risk of their own lives.'

William Scott reports that the pitmen want the day shift to begin as soon as they arrived at the pit rather than when they got to their positions, which could take up to an hour. They also complained that they were fined if they bought out small coals, stone or other materials, which was hard to avoid as they could see very little with the dim glow of the Davy lamp. The corfs that they sent up reflected this fact, and they were fined for all inappropriate material. In George's time 2*d* was the fine for foul coal, splent, stone and small coals. 3*d* was taken for a quart of bad material and double that for two quarts. If the weight of the unusable produce was four quarts then 5 shillings would be deducted from the man's wage with a legal penalty attributable to him. Rather than take it to the courts, coal owners would frequently penalise the miner 10 to 20 shillings. It would often be the case that the man's wage did not cover his fines, and he had spent a whole week working to incur a debt. The coal owners' attitude was 'that the fines are no hardship to a careful and honest workman, but merely operate as a due security against negligence and fraud'.

At West Moor Colliery in 1844, if the weight of a tub of coal sent up from the mine was 5¾ cwt rather than the designated 6 cwt, the miner was not paid but the coal was still taken by the owner. The miner had no way of knowing how much his tub of coal weighed at the bottom of the mine. At Fawdon colliery in the same year, the miners found the weighing machine to be inaccurate, meaning they each lost 4*s* 3*d* every week or £10 13*s* a year.

Community shops were run by coal owners in a system known as truck. These shops existed in the first place because of the isolation of collieries, but even when other shops sprang up the system stayed in place. It served the most deprived of the colliers: those who did not deal wisely with their money, those with young children or who had fallen on hard times. Colliers who were more financially prudent were not seen at the truck shop, but if a man needed an advance it was given on the requirement that it was spent at the company shop. If the money was not spent at the shop, the man was 'marked' and denied any more advances, sometimes even being sacked.

Even in the larger collieries, the shopkeepers were 'a little rough, some of them, at times in the way in which they treat their customers'. Women were not treated any better. In 1854, John Owen remarked that women could 'stay perhaps from seven in the morning until six in the evening until they could be served', and even in 1871 a secretary to the Truck Commissioners went to the Abersychan Company at 2 a.m. in November on 'turn book day' and discovered a group of women waiting for the shop to open even though it wasn't due to open for another four hours.

The miners had to work hard to gain respect. Even in 1872, a newspaper report on 22 June makes fun of the Durham miners in the light of their strike against butchers' prices:

> The butchers have already, I believe, virtually stopped their trade; and at the meeting, which is convened for Monday in the city of Durham, it will probably be resolved to reply to this novel sort of a strike by as novel a species of lock-out ... In short, the butchers talk of giving their customers a month's holiday, in which to make a grand experiment in vegetarianism ... The miners of Durham have very intelligent trade leaders; men who know perfectly well how market prices are governed, and who, in the conduct of their own relations with their employers, have of late been advocating a policy of temperance and mutual explanation. Would it not be well that this meat question were referred to the trade delegates, and that this body should meet with the butchers, and endeavour to come to a fair understanding? The butchers, I have no doubt, would be quite willing to make a fair statement of their profits, which might be compared with those which were made by tradesmen in other districts. If it were found that they

were asking only a fair market price for their meat, the uselessness and futility of the agitation would at once be made apparent.'

The alehouse was another place used to pay the men, although this was outlawed in 1842. The coal owners in fact backed this move, as they were fed up of losing hours due to hungover men not turning up to work, but it was another ten years before the rule was fully enforced.

At this time miners were charged in the region of £7 per annum for a house that was not worth more than £4 a year in rent. They were evicted without notice if they proved to be troublemakers and were sent packing, their furniture thrown out of their houses and their names circulated around neighbouring collieries to prevent them from getting further work. Pitmen even had to resort to changing their own name to hide their identity.

As concerns over safety grew, the attitude of the authorities was that prior to 1844 more lives had been lost at sea in a single shipwreck than had been lost in the previous fourteen years at the mines of Durham and Northumberland, with fewer mortalities than 1.5 in every 1,000 throughout that period. The pitmen replied that up to £15 million had been invested in the coalmines of Durham and Northumberland, employing at least 25,000 men, with most of the shipping trade of the Tyne, Wear and Tees engaged in the coal trade amounting to £3 million annually and generating thousands of jobs, to sailors directly and to countless other trades indirectly. The wealth of Great Britain depended on its coal. The pitmen kept scanty records of deaths in the mines of Durham and Northumberland during the years 1803 to 1843, but the evidence suggests 501 deaths between 1803 and 1820 from explosions, 75 from drowning, 9 from chokedamp, 18 from a burst boiler and 22 from other causes, reaching a total of 625 deaths. The years 1821 to 1830 saw a total of 370 deaths in the same collieries, 341 from explosions. The years 1831-1843 saw 451 deaths, 291 from explosions, leading to a total of 1,446 deaths overall in this time frame. This equates to 5.78 per cent of the workforce of 25,000, which is considerably greater than the 0.15 per cent suggested by the coal owners.

Tensions between the workforce and the coal owners grew. In 1815, a certain Mr Buddle was the manager at Heaton Main colliery and was notorious for exposing his workers to danger and getting them killed. On 3 May, Buddle wanted to access the coal pillars left in an old pit from an adjacent site. However, between them lay a dyke, an obstacle of stone or cinder coal, only a few yards wide but very deep. Where these are found, they change the level of the coal seams.

Mr Buddle gave his directions to cut through the dyke, troubling some of the miners who thought the old works were flooded. They wanted the dyke bored before they continued working, so that they could plug the dyke if they needed to. Mr Buddle did not want to listen to the knowledge of the workmen, relying heavily instead on his own scientific knowledge. The men were duly worried, wanting to quit, but they faced imprisonment if they did so. They discussed the matter and since it was decided it was a question of judgement, they would not alarm the other men working in the mine since most were working on a higher level. There were seventy-seven working in the mine that day.

They set to working on the dyke. It burst, drowning the miners immediately. It was to take nine months to clear the water. Some of the more experienced pitmen who had not been working that day spoke to Mr Buddle and said that some of the miners could have survived as they were working in the higher levels, which would still have access to oxygen, and urged him to try and rescue them. Mr Buddle did not listen to them and obstructed their efforts to go down one of the shafts to find them. After nine months, when they did eventually get down the shafts, their suspicions had been realised. The workers had not died of drowning; the bodies of two dead and butchered horses showed that they had died of starvation. They had eaten the bark from fir pillars, candles and everything else that they could get their hands on. Corpses of several boys had been placed in a deputy's store chest, as it was felt that they had died early, but chalk markings on the walls suggested that one particular miner, Ralph Witherington, had died much more recently.

Pitmen were very keen to have the safety side of mines properly looked at, with particular emphasis on ventilation. William Mitchell, a pitman in the Ouston colliery, wrote to the coal owners of Northumberland and Durham in 1844, saying that 'the lives of your petitioners are not a day nor an hour secure from such deadly operations, and, that it is a horrible and fearful thing to die such a death or live in daily expectation of it'. Colliers said that the only way to ensure adequate ventilation was by always sinking two shafts, and more if the acreage of the site demanded it. Mines in 1844 often covered 100 to 400 acres, some even above 500 acres, with 60 to 70 miles of underground passages. These proved very complex to properly ventilate, bringing the pitmen's lives into constant danger.

The pitmen were justifiably angry as they had no representation beyond their employers. The odds were very heavily stacked against them.

And that the fines must remain as before, as a necessary protection to the owners against 'negligence and fraud'. Now if these fines were only imposed for 'negligence and fraud', there might be some excuse for them; but they are often exacted for accidents, which the men are unable to prevent, and, in a very arbitrary manner, against which there is no appeal ... Thus coal-viewers are our accusers, judges, and executioners, and we have no mediators, or no other court to which we have access; but what they say and do are the laws by which we are coerced, and not the laws of our nation.

In 1841, a collier said, 'It is a life of great danger both for man and child: a collier is never safe after he is swung off to be let down the pit.' Twenty years later, an inspector observed that 'this account of death, contusions, fractures, amputations, and other surgical operations, altogether sounds like the description of military movements in the field rather than the ordinary report of industrious and peaceful pursuits'.

Prior to the Coal Mines Act of 1850, coal owners did not keep adequate records of injuries or deaths, and those in the trade did not like to detail these facts. There was a general lack of concern, shown by the informal course of action for inquests. At the inquest it was hard to get unprejudiced accounts and a decision that was free from fear or favour since testifiers and people belonging to the jury might be employees in the same mine and would not like to implicate their boss or colleagues. This ambivalence towards the miners was beginning to cease when, during the 1840s, there was an increased feeling that the mines should be under government regulation. This was initially a result of activities in the north of England, but a spate of serious disasters in South Wales brought about a transformation of attitudes. After a flood at the Landshipping colliery in Pembrokeshire killed forty people, journalists demanded government inspectors investigate mines and the proficiency of the people running them.

Two managers, Penrose and Evans, of the Welsh Eaglesbush and Eskyn collieries, took a case to court in 1845 against some of their workers who had abandoned the workplace without finishing their notice period. The magistrate hearing the case threw out the case, saying that the works had a terrible reputation for death from firedamp due to lack of ventilation, and therefore men working there were in great danger. The magistrate ruled that the men had a right to safeguard themselves, but additionally the supporters of the Neath Union should be protected from the expense of assisting the casualties and dead since neither the coal owners or workers had savings for a disaster.

An explosion occurred at Risca in Wales in 1846, with thirty-five men killed. These men had come from as far afield as Somerset and Gloucestershire; local men refused to work in the pit as they believed it to be dangerous. Two years later, twenty men died at the Eskyn colliery after the owners stopped using a furnace to create air flow and relied solely on natural ventilation to air the mine. The jury returned a verdict of accidental death, even though forty to fifty men had been burnt through explosions in the previous four years and there had been recent protests about safety at the pit. In 1849, fifty-two miners died in an explosion at Lletty Shenkin in the Aberdare Valley. Again, a verdict of accidental death was reached by the jury, although the government did send a witness to the inquest. Miners met after the hearing and decided to take five men from each mine to examine the workings and report on them. They found that the ventilation routes were not big enough, allowing dangerous gases to build up in the workings.

The growing sense of alarm was detected by the press, and it became clear that the government needed to be more involved. The Select Committee of the House of Lords scrutinised the separate colliery catastrophes in 1849, and passed the Inspection of Coal Mines Act in 1850 'to provide for the Inspection of Coal Mines in Great Britain'. It created a government authority on subterranean work and was the basis of the creation of safety regulations. It also recognised that it was not possible to put a stop to accidents occurring altogether, and so spearheaded the way to create a compensation scheme for the colliers.

The Act of 1850 required coal owners to retain precise maps of their mines and to inform the Home Office of every death that occurred. This Act also created four posts for inspectors who could legally go into any mine in Great Britain and inform the owners of any problems that they found. In some cases, if they found the colliery to be perilous, they could give legal notification that the mine was 'dangerous and defective'. Assessments started in November of 1850. The Act was intended to be temporary, with a duration of five years, but it was re-enacted in 1855.

It was harder to ensure safety in areas where production was expanding rapidly. This was especially true in the Aberdare Valley, where coal output grew from 477,208 tons in 1850 to 1,575,856 tons in 1860. Problems with gases were more likely to occur as existing pits were deepened and new ones excavated. The inspector's role was to put safety requirements first, even before business interests, and this did not sit well with the coal owners.

The Act and the role of the inspector lost momentum still further, as no legal penalties were enforced if the guidance of the inspector had been ignored. Lackadaisical juries would time after time give verdicts of accidental death, disregarding the proof, meaning that no criminal charges could be brought. On the whole, workmen were still unwilling to say anything in court against their managers or the coal owner, and those who did found it very difficult to gain subsequent employment. Very occasionally a verdict of manslaughter was achieved, but it was hard to find the right person to blame.

There were 738 fatalities in the South Wales and Monmouthshire mines in the period 1851–55. Explosions accounted for 173 of these deaths, roof collapses 300, shaft casualties 143, other causes underground 98, and 24 came about due to mishaps above ground. Boys aged from ten to fifteen years made up a ninth of the workforce but comprised a fifth of all fatalities. Many met their death through the 'crush of trams' as the weight of the vehicle was too much for them.

Benefits paid to the family after a fatality usually ended after a few weeks. Poverty was inevitable. The sympathies of the general public helped raise funds for major disasters and some coal owners such as George Insole and Thomas Powell added to these funds. Those who were injured were put forward for light duties and a widow could find support to pay for her house and coal until her children were older. Small accidents would not capture the public's notice.

The 1850 Inspection of Coal Mines Act meant that if a fault was detected at a colliery the owner had to pay £1 a day until it was mended, and any collier could leave his post without fear of reprisal until the fault was corrected. But the Act did not change the legal system that heard the cases of accidents at work, and all too often the magistrates were still severely biased towards the coal owners at the expense of the colliers. This drew comment from no less a figure than Lord John Campbell, later Lord Chief Justice:

> I really feel it is my duty to say that the magistrates had better consider whether an application of a criminal nature may not be made against them ... It is most shocking to think how a salutary Act for preserving the lives of Her Majesty's subjects engaged in mining has been thus rendered entirely nugatory.

The Mines Regulations Act of 1860, which was implemented by the following year, added to the list of safety rules. Fines for not imposing these measures, paid by owners, agents or viewers, were enlarged to £20. The age limit for operating the engine and shaft was raised to eighteen years. No boy was allowed to work unless they could read and write, or alternatively prove that they attended school for a minimum of three hours daily for two days a week. This effectively limited the minimum age at which boys could work in the mines to twelve years of age.

Other initiatives in this Act required the shaft to be properly lined, the cages to be covered, and sufficient timber to be supplied. Smaller collieries lacked the capital to buy up-to-date machinery, but even rich owners like Crawshay Bailey still used single-linked chains to lower and raise the cages in 1861. A collier named Thomas Evans refused to descend a pit belonging to Bailey, who was made to pay a £10 fine and promptly installed three-link chains at no small cost.

The next piece of legislation, the Coal Mines Act of 1862, prohibited the sinking of a new mine with fewer than two shafts. Older mines had to comply by 1 January 1865. In the Ferndale pit, 178 men died on 9 November 1867, followed by 53 deaths on 10 June 1869. These deaths were not due to bad ventilation; colliers had been tampering with their safety lamps, causing explosions. The manager of the pit explained, '[I] did not know what to do, they were a rough lot.' Although it was difficult to bring a prosecution, because men did not like to inform on others, several men were sentenced to two or three months in jail with hard labour.

Another Mines Regulation Act was passed in 1872, stipulating that mines with more than thirty employees must be managed by a certified person. The minimum age for a worker was still twelve, but hours had been cut to ten hours daily or fifty-five hours a week until boys were sixteen years old, with no night shifts. Two men were chosen from each pit to inspect the mines at least once a month. The 1860 Act had said that a coal owner was not permitted to sit in court when his own mine was in question, but this new Act of 1872 barred him from sitting in any circumstance. They were now liable, as well as managers and agents, to be put behind bars if safety issues were created by any 'personal act, personal default or personal negligence'. This was a huge step forward.

However, rates of change were slow. Often improvements took ten years to implement if they were made at all. An inspector called Mackworth observed, 'I have yet to learn that a collier or his widow can obtain redress for the wrongful act, neglect or default, which breaks down his health, cripples him for life, or condemns him to a violent and needless death.' The Fatal Accidents Act, which ruled that relatives could claim compensation against a coal owner on behalf of any miner that died at work, had been brought in by Lord Campbell as long ago as 1846, but it had proved unworkable. Solicitors were generally unwilling to take on such cases, and even if they did the cost of the action against such a powerful opponent made it impossible. It was not until 1875 that an accident fund was created in South Wales, even though discussions had begun after the explosion at Cymmer colliery in 1856, when 114 men and boys were killed.

Nonetheless, slow progress was being made. In the seven years up to 31 December 1862, there had been one fatality for every 42,421 tons of coal produced. By 1875, this had improved to one fatality for every 88,890 tons of coal produced in the past three years. The influence of the government inspectors really made a difference to safety standards. Although they were unpopular to start with, they became well respected.

2

THE INDUSTRIAL
REVOLUTION

The Industrial Revolution began in the eighteenth century, with Great Britain leading the way due to its dominant colonial power and huge overseas trading empire, most notably in the form of the East India Company. The ability to trade overseas was a leading factor in bringing about the Industrial Revolution. Historians concur that this period wrought the greatest changes for mankind since the domestication of animals and the cultivation of plants.

The nineteenth century brought great wealth for Britain, which held 9 per cent of the global GDP in 1870 while its empire as a whole held a massive 24 per cent. The technological shift towards machinery came between 1760 and 1840, with steam power, machine tools and factories replacing hand production and cottage industries. Textile manufacture was the cornerstone of the Industrial Revolution, with automated spinning taking off as early as the 1780s. By the 1830s and 1840s the initial innovations of the previous century were well entrenched, and another burst of development arrived in the 1870s. This is sometimes referred to as the Second Industrial Revolution, when trains and steamships came about, along with the invention so important to George Elliot: the electrical telegraph.

The 1830s witnessed the emergence of an automated cotton spinner, fuelled by water or steam, which escalated the production of an employee by a factor of 500. The power loom increased production by a factor of 40, while the cotton gin (which separates cotton fibres from their seeds) raised output by a factor of 50. The wool and linen industries also saw growth.

In 1788, there were 50,000 spindles in the United Kingdom. By 1820, this had increased to 7 million. In 1750, 2.5 million lbs of unworked cotton had been brought into the UK for spinning and weaving, mostly by a cottage industry; within forty years, importations had risen to 22 million lbs, predominantly for machine working. Some 52 million lbs of cotton was imported in 1800, and the middle of the nineteenth century this number had rocketed up to 588 million lbs.

The growth of the cotton trade took it from 2.6 per cent of the UK economy in 1760, with one-third exported, to 22 per cent by 1831, when more than two-thirds left the country. Wool, meanwhile, was worth 14.1 per cent of the national economy by the turn of the nineteenth century. Prior to this time, India could produce cotton garments at a much more inexpensive rate as labour was considerably cheaper than in the UK. This had all changed with the advent of James Hargreaves's spinning jenny, patented in 1770.

Iron was now made using coal rather than wood. Coal required less labour to extract, and wood, which was in short supply, needed to be heated to create charcoal. The coal-driven practice of puddling, developed in 1784 by Henry Cort, used the reverberatory furnace to decarburise iron, giving it a higher melting point and allowing it to be used for construction at a comparatively low price. However, puddling involved the hand-operated stirring of the iron using a large rod. This was strenuous work, and most puddlers did not live to see forty. The procedure was refined by Baldwyn Rogers in 1818 and further by John Hall in 1838. It continued until the late nineteenth century, when iron was finally replaced by steel. Ironworking businesses evolved quickly with the availability of this cheaper material. The most obvious result was the growth of machine tools, such as the engine lathe, which aided accuracy and ease of production, enabling bigger, stronger engines to be built.

Numerous other inventions emerged or were refined at this time. James Watt developed his famous steam engine, which worked in a rotating fashion rather than by the reciprocating method, allowing its use for pumping in commercial operations. The stationary steam engine was an essential element for industry, enabling Britain to increase its steam horsepower from 10,000 in 1800 to 210,000 by 1815. Humphry Davy invented his eponymous lamp in 1815. A few years later, Michael Faraday had developed the first electric motor. Samuel Morse created the first electric telegraph in 1844, allowing communication over great distances through use of a wire. Edison's light bulb was pioneered in 1879, lasting for forty hours. A year later, he had developed a bulb that lasted for 1,200 hours. Similarly named and timed but separate, the British Agricultural Revolution started in the mid-1700s and carried on until the late 1800s, bringing about increased land and labour productivity. The invention of the seed drill by Jethro Tull in 1701 precipitated this period, although his invention was initially expensive and erratic, and good seed drills were not available until past the midway point of the eighteenth century. The threshing machine developed by Andrew Meikle in 1784 displaced many farm labourers as only a quarter of them were needed. These redundant agricultural workers were on the brink of starvation, resulting in the Swing Riots of 1830, which began when threshing machines were destroyed in the Elham valley of east Kent, causing violent public disturbances across the whole of southern England and East Anglia. By the late nineteenth century, machine tools developed thanks to new metalworking processes led to the mass production of farm machinery, among them the reaper, binders and combine harvesters.

Coalmining in South Wales had been around since the Roman invasion, but it did not really get going until the seventeenth century with the advent of copper extractions. Before the use of the steam engine, shallow ditches called bell pits were created to extract coal from seams just below the surface. In hilly areas, drift mines were plunged into hillsides. Shaft mining had its problems as water often collected in the depths and buckets of water had to be used to remove it. Thomas Savery invented the steam pump in 1698 and the Newcomen steam engine followed in 1712, both enabling the elimination of water. They were costly machines, generating 5 horsepower, but they caused a great proliferation in the depth of mining, allowing more coal to be extracted. They were used in the coalfield until the first few decades of the nineteenth century.

Richard Trevithick and Oliver Evans, at about the same time, developed a higher-pressure, non-condensing steam engine which had an engine and boiler small enough to operate on the road, rail and on steamboats. John Smeaton and James Watt improved it from 1770 onwards, lowering fuel expenses and making the mines more lucrative. The Cornish engine, invented in 1810, was a further improvement on the Watt steam engine.

Prior to the Industrial Revolution, goods were transported by rivers, road and sea. Although wagonways were used to get coal to rivers, canals had yet to be established. Horses and sail were still the dominant transport modes, with horse railways introduced in the late 1700s. Steam trains came about in the first few decades of the 1800s, and new equipment raised sailing speeds by 50 per cent from 1750 to 1830. The Industrial Revolution saw Britain's inland transportation network vastly upgraded, with turnpike roads, canals and waterway systems allowing goods to be moved more rapidly and at a lower price. This new system also facilitated the spread of ideas.

Rivers were straightened, widened and deepened and impediments cleared. Locks were developed. By 1750, Britain possessed more than 1,000 miles of waterways to be used for transporting goods. Horses were able to pull freight on a barge twelve times heavier than what they could pull on a road. The construction of the Bridgewater Canal in the north-west of England in 1761 cost £168,000 and was paid for by the 3rd Duke of Bridgewater, who wanted to take his coal from mines in Worsley to Manchester. He saved so much money in transport costs that the value of coal halved in Manchester within a year of the canal's completion. Many canals were swiftly constructed to replicate this success, in particular the Leeds and Liverpool Canal in 1774, and the Thames and Severn Canal in 1789. By the 1820s a countrywide infrastructure was in place and many lessons had been learnt that would be brought to bear in the construction of the country's rail network. The latter became a cost-effective option by the 1840s and soon grew to dominate; the last great canal constructed in Britain was the Manchester Ship Canal in 1894, establishing that city as an important port. However, its failure to live up to expectations cemented the primacy of the railways, which proved faster and more economical.

Three inventions enabled the building of railways. First was the development of cheap puddled iron, which came about after 1800. Then there was the rolling mill, which forged the rails, and finally the high-pressure steam engine. Boulton and Watt's patent of their high-pressure steam engine expired in 1800, which allowed other inventors to climb on the bandwagon. Watt had understood that engines spent a huge amount of energy continually cooling and reheating the cylinder, so he redesigned it, giving it a separate condenser, making the engine more powerful, efficient and cost effective.

Before the Industrial Revolution, most of the population worked in agriculture, either as tenant farmers or landless labourers. Families would spin and weave their own garments, providing surplus for sale. Merchant capitalists provided the raw materials and paid the families for each piece. Globally, the market for cotton goods was dominated by India, China, some parts of Iraq and to a lesser extent the rest of Asia and the Middle East, whereas wool and linen were generally popular in Europe. By 1792, it was possible to purchase a forty-spindle spinning jenny for £6, which most cottagers could afford. Larger machines came onto the market, however, including spinning frames, mules and power looms, and these were unobtainable for cottagers, particularly if they were propelled by water. This marked the beginning of the capitalist system of factories.

Factory workers were predominantly unmarried women and children, especially orphans. They worked a twelve-to-fourteen-hour shift every day except Sunday. Women frequently worked seasonally, appearing only when there was no farming to do, but it was unpopular work – the hours were long, the pay was bad and travelling to work was hard. Many people worked in the factories as a last resort.

The major increases in productivity did not necessarily equate to better living conditions. Real wages only grew by 15 per cent from the 1780s to the 1850s, while life expectancy was effectively static until the 1870s. However, children were better off. Infant mortality rates in London dropped from 74.6 per cent in 1730–49 to 31.7 per cent in 1810–29.

Food was a luxury for most of the world until the late 1800s, with malnourishment the norm, and Britain was no exception. Up to 1750, life expectancy in Britain was around forty years, with France even lower at thirty-five. In the US, where food was more abundant, people could expect to live for around forty-five to fifty years.

However, more advanced technology did not mean lower food prices. Britain and the Netherlands had advanced agricultural processes at the beginning of the Industrial Revolution and created more food, but their populations grew as well. Thomas Malthus in his 1798 book *An Essay on the Principle of Population* describes an increase in food leading to population growth and a resultant fall in standards of living, which became known as the Malthusian trap. The Corn Laws, whereby taxes were levied on imported food and corn from 1815 to 1846 so that domestic producers could charge higher prices, greatly affected resources in Britain. The Great Irish Famine of 1845 to 1849, in which a million people died and another million emigrated, reducing Ireland's population by a quarter, saw the repeal of the Corn Laws.

Populations in the newly industrialised cities increased tenfold between 1801 and 1901, whilst the total British population saw a fourfold increase. Financing the increased city housing fell on the building societies, which dealt directly with large building companies. Most people rented, but it was difficult to accommodate everyone so recent arrivals on low incomes were forced into increasingly congested slums. There they had to endure appalling conditions, with a lack of safe water, sanitation and health provisions. Death rates were high, particularly in children and young people, and tuberculosis was rife. Cholera and typhoid were everywhere.

Much was written about this state of poverty, for example by Charles Dickens in the mid-nineteenth century. In *The Condition of the Working Class in England* (1845) Friedrich Engels describes the slums of Manchester, where people dwelt in makeshift shacks that exposed them to the elements, with muddy floors, no washing facilities and an incredibly high density of people. He writes of the terrible smells of untreated sewage and how the rivers flowed green. In 1854, John Snow managed to trace a cholera epidemic in Soho to the faecal leakage of a home cesspit into a public water supply, correctly deducing that cholera proliferated through contaminated water. Although his hypothesis took several years to gain acceptance, it eventually brought about crucial transformations in the water and sewage systems in London and other British cities as well as inspiring changes in public health around the world. Throughout the nineteenth century there were parliamentary Acts standardising sanitation and house building among other things, leading to Engels commenting in the 1892 edition to his book that matters had been substantially bettered. One of these Acts was the Public Health Act 1875, which regulated housing and ensured that a privy was established in each backyard with municipal collection of night soil (human waste).

Consumers could afford more after the Industrial Revolution. The cost of clothing fell, as well as household commodities like cast-iron cooking products. As time went by, people were able to purchase cooking stoves and items such as tea, coffee, sugar, tobacco and chocolate came within reach of the average person. For those with a higher income, entrepreneurs such as Josiah Wedgwood began producing fine china and porcelain.

The Industrial Revolution marked the first time in history where a population explosion was met with an increased individual income. England's population rose from 8.2 million in 1801 to 16.7 million in 1850. By 1901 it had risen to 30.4 million. Europe itself saw an increase from 100 million to 400 million between 1700 and 1900. This was undoubtedly due to better living conditions.

The newly created middle class of industrialists and businessmen overpowered the historic landed class of aristocracy. The working classes had many opportunities available to them in the contemporary factories and mills; however, the new technology set the pace, and shifts were long and uncompromising, although these constraints had also been present in work before the Industrial Revolution. Factories created jobs and people flocked to the city to get them. Manchester was known as 'Cottonopolis', being the first industrial city in the world, and its population grew by six times between 1771 and 1831. Bradford saw an increase of 50 per cent in its population every decade from 1811 to 1851.

There was also mass migration, with 20 per cent of Europeans moving abroad between 1815 and 1939, mainly heading for the United States. They left for the same reasons: the loss of peasant farming and cottage industry as a source of income, poverty and a rapidly growing population. They were enticed by the pull of work opportunities, obtainable land and affordable transportation. However, not all found their pot of gold, and many returned home. Demographically, in 1800, less than 1 per cent of the world population was made up of immigrant Europeans and their offspring. By 1930, that number had risen to 11 per cent.

An exploding population meant more children. And for the children, education was generally superseded by the need to work.

A child's pay was much lower than an adult, even though they were capable of achieving as much; the new machines required little brawn to operate, and as they were so recent none of the grown-ups had much expertise. This made children very popular among employers. Indeed, in 1788, two-thirds of workers in the 144 water-powered cotton mills in England and Scotland were children.

Children were paid between a tenth and a fifth of what a man would earn and were often employed from the age of four. They worked long hours and were not well treated, often being beaten. Children in the mines would frequently get lung cancer and other ailments and many did not live to see their twenty-fifth birthday. Children from the workhouses were sold, or abandoned as pauper apprentices, and would be forced to toil long hours with no pay, just to earn their board and lodgings. If they tried to escape, they would be flogged and handed back to their employers. Some were even chained so they could no longer flee. In the cotton mills, children would work as scavengers and had to creep under the apparatus to collect cotton, with the ever-present danger of losing a hand or limb – some were even decapitated. Others were crushed by the appliances. Work was long and hard; children often worked for fourteen hours a day, six days a week. Small girls were commonly employed at match factories where they suffered from phossy jaw, a condition in which the bones of the jaw were destroyed by the white phosphorous present in matches. Other industries had their dangers too; at the glassworks young children were frequently blinded or burned, and in potteries children would come into contact with deadly clay dust.

Investigations into the working conditions in the mines and textile factories shocked the upper and middle classes and helped bring about changes for the children. Politicians tried to amend the laws, but they were up against factory owners who claimed to be helping the children out of starvation by employing them. It was in 1833 and 1844 that the government passed the Factory Acts, the beginning of the prohibition of child labour. Laws were passed forbidding children under ten from working at all, and older children could only work twelve hours a day and never during the night. Factory inspectors operated to uphold these laws, but there were not enough of them to make much difference. Child labour continued to be a considerable part of the workforce in Europe and the United States up into the twentieth century.

Since the Industrial Revolution collected people together to work in mines, mills and factories, it also saw the establishment of combinations or trade unions, which aimed to support the workers. These unions had the power to stop production completely with a view to request better working conditions. Employers had to decide whether to cave into the unions at a cost to themselves, or suffer the expense caused by loss of production. Unions were more successful where their workers were skilled, and could not be replaced easily.

Strikes were difficult for both owners and workers. The Combination Acts of 1799 banned employees from joining a union. This was rescinded in 1824, although unions continued to be seriously constricted. Herbert Evatt, in his book *The Tolpuddle Martyrs*, describes how one British newspaper reported that unions were 'the most dangerous institutions that were ever permitted to take root, under the shelter of law, in any country'.

The Friendly Society of Agricultural Labourers was formed in Tolpuddle, Dorset, in 1832 with the intention of dissenting over the issue of their falling wage. The men would not work for less than 10 shillings a week, even though their wages had been cut to 7 shillings and were due to fall to 6 shillings. Two years later, in 1834, resident landowner James Frampton corresponded with the Prime Minister, Lord Melbourne, referencing a hazy law from 1797 which forbade citizens from making sworn statements to one another. Six men – James Hammett, George Loveless, his brother James Loveless and brother-in-law Thomas Standfield, Thomas's son John Standfield as well as James Brine – were subsequently arrested, charged and transported to Australia. They are well known to this day as the Tolpuddle Martyrs.

The Luddite movement began with the lace and hosiery workers in the Nottingham area and gathered pace in other aspects of textile production. The use of more productive machines, which needed very few labourers, and unskilled ones at that, meant that many people were left without a job. By 1811, events had turned violent, as some people attacked and destroyed the machines and factories which had rendered them jobless. They were named after Ned Ludd, who in 1779 was thought to have destroyed two stocking frames in an angry outburst. His identity was seized upon, and he became a folk hero. The Luddites quickly found favour and the government was forced to protect businesses with the threat of the armed forces employed to restore order. Anyone caught in the act of destroying factories or machinery was sentenced to death by hanging or transportation for life.

The Swing Riots, led by agricultural workers in the 1830s in large swaths of the south of England, claimed a fictional character as their leader. This was Captain Swing, and aggressive letters were sent to landowners under his name. He was depicted as a conscientious tenant farmer who had lost his job due to the introduction of new threshing machines and was living in a state of poverty and ruin. The rioters attacked threshing machines and set fire to hayricks. They were successful in that their movement led to the beginnings of the trade unions, which created momentum for change.

The Chartist movement of the 1830s and 1840s was the earliest, most extensive working-class political group to demand constitutional equality and civic egalitarianism. An enormous 3 million signatures were collected on their charter of changes, but it was disregarded by Parliament without deliberation.

Gradually unions were able to change the law to enable them to strike legally. The year of the general strike was 1842, and involving cotton workers and miners across the whole of Great Britain stopping output on a wide scale. The right to vote was increased more widely in 1867 and 1885. Socialist political parties were beginning to emerge with the support of the trade unions, allowing working people an authoritative voice. These parties in time would amalgamate, creating the Labour Party in 1900.

The entire globe was affected by the beginnings of the Industrial Revolution in Britain. Much of Asia historically relied on handmade textile manufacture for income, and these nations could not keep up with the machine-made fabrics coming from Britain. Over a few decades their business was obliterated, leaving millions unemployed. Many of them starved.

This new manufacturing ability divided the world financially in a way that had never been seen before. In 1750, Europe was responsible for just under a quarter of the total global manufacturing output, with the US producing 1 per cent, Japan just under 4 per cent and the rest of the world 73 per cent. Over the next 150 years this was to change completely. The main shift took place from 1800 to 1860, as Europe went from producing 28.2 per cent of the goods to a little over half in 1860. In the same period, the share produced by the rest of the world fell from 67.7 per cent to 36.6 per cent.

The emergence of the cotton gin in 1792 allowed huge cotton plantations to develop across the United States and Brazil. Out of a global output of 490 million lbs of cotton in 1791, the US contributed just 2 million. By 1800, US output had risen to 35 million lbs, roughly half of it for export. By the twentieth century the US would be responsible for almost all cotton production.

The Americas, especially the US, had a different problem to Britain in that they struggled with a lack of workers, who subsequently charged an inflated price for their time. Their response was to resort to the slave trade. Cotton plantations were extremely profitable, and this was one industry which could meet the market demand. However, the US Civil War of 1861 to 1865, in which slavery played a large part, led to a cotton famine.

Environmental legislation was first passed in the mid-1800s, resulting from increased air pollution. Enormous factories, and the accompanying rise in coal use, led to rocketing levels of smoke contamination in manufacturing towns and cities. Chemical pollution became a problem after 1900, with an increase in toxic substances, as well as the vast level of sewage from an enlarged population.

The first environmental laws, the Alkali Acts, were passed in 1863 and focused on reducing harmful gaseous hydrochloric acid, a byproduct of the Leblanc process, which was used to make soda ash. The government assigned an alkali inspector, with a team of four, to clamp down on these emissions. As time went on, the inspectors gained greater powers until 1958, when the Alkali Order regulated all significant businesses that gave off smoke, dust, grit and fumes.

The gas industry began in the UK in the 1810s, producing extremely poisonous liquid waste that was fed into waterways. Companies were taken to court and habitually lost, so they adapted their most harmful activities. Companies in London were frequently prosecuted in the 1820s for contaminating the River Thames and killing its fish.

The Public Health Act of 1875 stipulated that all fires and furnaces must be designed to absorb their own fumes, penalising factories that emitted considerable smog. The powers of this Act were again increased in 1926, with the Smoke Abatement Act encompassing other discharges, such as ash, soot and grit, and gave local councils increased power to enforce their own laws.

After 1890, local experts took steps to encourage community action, determining environmental problems, and working to initiate change. Water and air pollution usually took precedence. NGOs were taking shape – the Coal Smoke Abatement Society was created as early as 1898 by Sir William Blake Richmond.

Industrialisation in mainland Europe took place later than it did in the United Kingdom. Many industries took the mechanisms that had been developed in Britain, buying them or learning them from British engineers and businessmen who had moved abroad. Belgium was the second country in Europe to industrialise, led by the Englishman John Cockerill. Germany would become a world leader in the chemical industry in the late 1800s; progress there had been hampered by their fractious state, and it was not until unification in 1870 that growth increased, especially in their railway network. The French Revolution and Napoleonic Wars (1789–1815) prevented industrialisation taking hold in France. Sweden was the first nation in the world to introduce compulsory schooling free of charge in 1842. Between 1850 and 1890, the Scandinavian nation experienced a burgeoning rate of exports, from crops to steel and wood, as they did away with obstacles to free trade in the 1850s.

The Industrial Revolution did not hit Japan until 1870 when Emperor Meiji modernised feudal Japanese society, constructing railways, revamping roads and building a Western-style education system. Over 3,000 Western teachers were employed to teach in Japanese schools, and thousands of pupils went to Europe and the USA to learn new skills. Politicians were also sent abroad, allowing Japan to move into the Industrial Revolution more quickly. Textile industries such as cotton and silk were among the first to be developed.

The USA was an agricultural economy, relying heavily on its natural resources until the late 1800s and early 1900s. Railways and waterways carrying steamboats were initially very important to move goods across this huge and sparsely populated nation. In terms of contributions to industry, they invented the cotton gin and the milling machine.

One issue due to the rise of the United States was what is now called brain drain. In Britain in the late 1700s there were laws in place to stop the emigration of skilled workers, especially from 1780 to 1790, when independent colonies were flourishing and the government was determined to prevent British knowledge and ability falling into competitors' hands. In 1789, Samuel Slater broke away. He travelled to New York, hoping to cash in on his skills, and founded fourteen textile mills. However, with a growing population and not enough work, ministers began to feel the pressure. They knew that without a relaxation of emigration policy there would be social unrest. In 1860, more than two-thirds of the immigrants to the USA and Canada departed from Liverpool.

Historians have put Britain's place at the forefront of the Industrial Revolution down to several key factors, among them the harmonious unification of Scotland and England; the lack of trade tariffs between the two countries; laws imposing rights of property and honouring contracts; a simple legal framework which permitted the development of corporations; the withdrawal of tolls (in the 1400s); and finally, but by no means least, a free capitalist market. Historian Jeremy Black sums up the Industrial Revolution:

An unprecedented explosion of new ideas, and new technological inventions, transformed our use of energy, creating an increasingly industrial and urbanised country. Roads, railways and canals were built. Great cities appeared. Scores of factories and mills sprang up. Out landscape would never be the same again. It was a revolution that transformed not only the country, but the world itself.

Britain's status as an island, with considerable coastal areas and traversable streams, at a time when water was the predominant method of moving goods helped its position enormously. Equally beneficial was the fact that Britain had the best coal sources in Europe. But the spirit of the age was also created by self-serving, entrepreneurial individuals who amassed great riches while allowing cheaper goods to be bought by all – in a nutshell, the consumer revolution. Life got better in Britain for most people, and on observing these benefits other countries became keen to follow suit.

Again, it was Britain's island status that helped it prosper as the nation was shielded from the plundering of the Napoleonic Wars that financially devastated Europe. Battles took place overseas. The British had obliterated European merchant vessels and were alone in having an effective fleet, allowing greater trade opportunities. Countries such as China and Russia were governed by more powerful sovereigns and lacked the flexibility to adapt and quickly exploit the new scientific and technological advancements. Added to this was the abundant natural resources in Britain. Coal, iron, lead, tin, copper, limestone and water were all readily available in the north of England, the Midlands, South Wales and the Scottish Lowlands, allowing industry to flourish. Cotton is best spun in moist, temperate regions – little wonder that the industry began in the north-west of England.

Religious tolerance in the UK also played a part. Voltaire wrote in his *Letters on the English* in 1733,

Take a view of the Royal Exchange in London, a place more venerable than many courts of justice, where the representatives of all nations meet for the benefit of mankind. There the Jew, the Mahometan [Muslim], and the Christian transact together, as though they all professed the same religion, and give the name of infidel to none but bankrupts. There the Presbyterian confides in the Anabaptist, and the Churchman depends on the Quaker's word. If one religion only were allowed in England, the Government would very possibly become arbitrary; if there were but two, the people would cut one another's throats; but as there are such a multitude, they all live happily and in peace.

There were many who abhorred the newly industrialised country and made their views clear in literature and art. Among them was William Blake, who cherished nature as opposed to industry:

And did the Countenance Divine,
Shine forth upon our clouded hills?
And was Jerusalem builded here,
Among these dark Satanic Mills?

All of these circumstances combined to create a lively world of burgeoning commerce and technological advancement, and it was into this fertile environment that young George Elliot arrived, ready to take his place in history.

3

GEORGE GETS GOING!

Even as a boy, living in Shiney Row, George Elliot wanted to get on in life. He taught himself the fundamentals of mathematics and surveying, and began to fund his education whilst still very young, setting aside a quarter of the wages he made from his work in the pits. As his granddaughter Florence recalls:

The picture vividly drawn in these lines was not exaggerated; such was actually the lot of George Elliot and many more like him. But the spirit that bought him success was inborn. Even in the gloomy recesses of the subterranean world, while his tender limbs ached under the strain of uncongenial labour, his mind was set upon better things, and he looked forward to a higher and brighter life in the upper air, away from the dim shadows and the grime of the place beneath. This could be achieved only 'by labour and intent study'. His hours at home, few as they were, were occupied in the task of self-culture. As he grew towards manhood, George Elliot mastered the practical operations of mining, and at the age of seventeen he placed himself under the tuition of two well-known mathematicians, George Watson and Peter Nicholson, to whose kindly and timely assistance he has attributed much of his subsequent prosperity.

In 1831, aged just sixteen, George was apprenticed to land surveyor and mining engineer Thomas Sopwith, who later worked with Isambard Kingdom Brunel and George and Robert Stephenson. This apprenticeship was George's first step to fortune. He and Sopwith investigated the coal supplies in the Forest of Dean, also surveying the proposed route of the Great North of England Railway linking Darlington and York. The line opened on 4 January 1841, with the Newcastle and Durham Railway following on 18 June 1844. A train could leave Euston at 5 a.m. and arrive at Gateshead in six hours and forty-five minutes, running at a speed of 45 miles per hour over the 303 miles. The young man's career was surmised thusly by his granddaughter Florence:

> The young man's application, and his desire to get on, were recognised by those above him in authority; and, when he was about eighteen years of age, he went, with the sanction of the viewer of Penshaw Colliery, into the offices for a few months, after which he resumed the ordinary pit work at the colliery. At 20 years of age he became deputy, and a year after that was appointed overman. Speedy promotion awaited him, for the following year found him officiating as under-viewer at Monkwearmouth, then the deepest colliery existing. Two years later he was appointed head viewer and continued in that office until the year 1851.

> Mr Elliot had achieved considerable eminence as a mining engineer, and his services in that capacity were greatly in request in the district. In 1840, Messrs. Backhouse and Mounsey on Mr. Elliot's advice, purchased the Washington Colliery from Mr. Russell of Brancepeth Castle, with the original owner of the celebrated Wallsend Colliery. Mr. Elliot took equal shares with Messrs. Backhouse and Mounsey, and this was the beginning of his career as a colliery owner. He undertook the personal management of the colliery for himself and his partners. Three years later, in 1843, Mr. Elliot took a lease of the mines at Usworth, which, although at the outset unremunerative, were, by his skill and perseverance,

ultimately greatly extended and improved. He sank pits from which a thousand tons of coal were raised daily, and after some years the lower seam was reached, and proved to be gas coal of very high quality, giving assurance of the success of the investment. In 1851, Mr. Elliot relinquished his position of head-viewer at Monkwearmouth and was appointed chief mining engineer to the late Marquis of Londonderry – a position which he held for many years, until his multifarious duties made his resignation necessary.

About the year 1864, he became owner of Penshaw Colliery, where forty years previously, his working life had been inaugurated, and he took up his residence at Houghton Hall. Penetrating also into Wales, he became purchaser of the Powell-Duffryn collieries, to which he subsequently added Aberaman, Cwm Noel, and other properties, making in the aggregate a coalfield capable of producing 6,000 tons daily. In this undertaking he was joined by some friends, including Messrs. Brassey, Swift, and others. Owing to his connection with Wales, he was made Deputy-Lieutenant and magistrate for Glamorganshire and Deputy-Lieutenant for Monmouthshire. In South Wales, also, Mr. Elliot became connected with the Alexandra Docks at Newport, Mon., whose ancient port had suffered from torpid indifference until the deflection of its trade to the neighbouring docks at Cardiff, erected by the enterprise of the Marquis of Bute. The Alexandra Docks were a great success, the advantageous geographical position of Newport having been demonstrated beyond a doubt.

For years George used his utmost endeavours to ameliorate the condition of the miner. It was his boast that he had preserved a table on which he and Lord Aberdare had drawn up an agreement that led to the introduction of the Mines Regulation Act, whereby the hours of daily labour were reduced from fourteen to nine. He was a member of the royal commissions on coal supply and on accidents in mines. His signature to the report of the commission, issued in 1871, was qualified by the observation that he did not approve of the introduction of one of Professor William Stanley Jevons' calculations, which seemed to imply the possibility of the exhaustion of British coalfields in one hundred and ten years. At the request of his fellow commissioners, he undertook an investigation into the quantity of available coal in the fields of County Durham, which in the result he estimated at upwards of 6.2 billion tons, and he subsequently calculated, as the result of another inquiry, that the coal to be found at all depths in North Staffordshire amounted to more than 4.8 billion tons.

In 1874 he was created a baronet in recognition of his public services:

> Sir George Elliot was a member of the Royal Commissions on Coal Supply and on Accidents in Mines, and in 1874, in recognition of his public services, he was created a Baronet. He was elected a member of the Iron and Steel Institute in 1877. He was also a member of the Institution of Civil Engineers and of the Institution of Mechanical Engineers.

Writing in 1945, the Technical Advisory Committee on Coal Mining looked back to George's era in the coalfield:

> The years of which we write ... were the days of the pioneer ... When we come, therefore, as we must, to point out the mistakes which were made in these early years of the coalmining industry, let us beware of merely being wise after the event, or of with-holding the need of praise due to a great race of men, employers, mining engineers, workmen and machinery workers alike. For whatever their faults, they were fit to rank with the greatest of Britain's industrial pioneers.

In the mid-eighteenth century, skilled men from the north-east were beginning to make their way into the South Wales coalfields, bringing new engineering and mining abilities. Among them were George Elliot and George Wilkinson – the latter becoming the first manager of the Powell Duffryn Steam Coal Company. They found that mining in Wales was different from other areas, as inroads into the coal, rather than standing firm, were liable to contract. This was due to the soft earth in Wales, which caused the land to close up.

Ventilation was improved by bringing in these men from the north, who advanced techniques. If a mine was left unworked for twenty-four hours, it would infuse with firedamp. Innovations were sorely needed as the industry grew, especially with the opening up of the Aberdare Valley, which contained hazardous coal lines. Parliament scrutinised mining fatalities between 1849 and 1854, from which arose a quite considerable difference in opinion as to which measures to take. South Wales became a fertile ground for new ideas. Although advances in ventilators and fans found favour, by 1858 the emphasis was still on furnace ventilation. It would be another ten years before mechanical means of ventilation began to be seen as the norm.

Another technical challenge was moving the coal. In the early days of 1840, the work had been performed by boys, girls, women and horses, but this was costly. Pit owners were realising that introducing tramroads on the levels and using horses to pull a cart would save the cost of five employees, amounting to £150 per annum. The Blaenavon Iron and Coal Company paid £14,000 per year for the use of 300 horses, to which was added the cost of a haulier and a boy to look after the animals. Roughly a third of the horses retired or died each year. The total cost for each horse and the associated employees and maintenance was £125 per year.

Mechanical haulage was therefore a popular alternative. This took the form of either the endless rope or the tail rope. South Wales predominantly used the tail rope as it was expensive to use the endless rope, which needed double tracks. Dowlais colliery first used the tail rope in 1861, transporting up to thirty-five trams, equating to 36 tons of coal, in one journey. The Powell Duffryn Steam Coal Company had relinquished their pit ponies by 1872, and all its collieries were using machines within the next three years.

Originally, steam power or a water balance was used to bring the coal from the pit eye to the surface. The water balance, holding about 2 tons of water, would be lowered to the bottom of the shaft, in balance with a tub containing a ton of coal, which would be raised as the water lowered. Long, dry summers could impede work, and steam winding was used in the deeper pits. Technicalities in coal-cutting machines were still on the drawing board in 1875. South Wales was a particularly difficult area to mine as roofs were unstable and seams were not even.

Although the output per miner increased in South Wales from the year 1854 (250 tons) to 1881 (343 tons), output compared to the rest of the UK was considerably lower. This is explained by the large degree of anthracite coal in parts of South Wales, but also exhibits the tricky environment in the South Wales coalfield. In the beginning of the 1870s, the length of the miner's day was also changing, with a nine- or -ten-hour shift in the week and a half day on Saturday.

Thomas Powell, a world-renowned coal owner, began life as a timber dealer, working in Newport. His father died when he was fourteen years old, leaving him in charge of a small timber yard that did not really cover costs for Thomas and his mother, as described in the *International Directory of Company Histories*:

> Powell was quick to realise that the introduction of new steam-driven engines and machinery would bring a rising demand for a more efficient and powerful fuel source. Although scarcely in use at the time, coal was to become the power source of the Industrial Revolution. Powell recognised that the coal-rich Welsh valleys offered an opportunity for a vast fortune...Powell, and the company he founded, profoundly transformed the South Wales area from, for the most part, rural, agricultural, and pasture lands, into what one source described as a countryside where 'ravaged valleys were

bestrode by giant bridges, great waterways and railways, and roaring towns with brutalised and degraded inhabitants.'

By 1830, Powell was considered a successful businessman. But he was soon to become more than this. By the end of that decade, Powell joined the ranks of an exclusive few. The Industrial Revolution had brought a new breed of men to the forefront of society. Known to some as tycoons and to others as tyrants, these men controlled the fuel and materials that served as the backbone of the Industrial Revolution. Some, like Powell, became infamous for their unscrupulous pursuit of wealth and power.

Powell began his interest in the coal trade in 1810, and by 1841 he owned four mines in Monmouthshire and Glamorgan, working with Thomas Prothero and Buttery Hatch, engaging a total of 431 men and boys. His mines were prolific in their extraction of coal, transporting a quarter of all the coal carried on the Glamorganshire Canal in 1840 – some 62,130 tons. However, at the time of the Children's Employment Commission of 1842, most mines were on a much smaller scale, employing fewer than 150 people. Thomas Powell was exceptional in his expansion of business, opening more and more collieries in a greater number of valleys.

Powell was quick to understand the growing trade in steam coal from Cardiff, working the untouched Aberdare Valley and expeditiously opening further mines to serve this market: Old Duffryn was sunk in 1840–2; Upper Duffryn 1844–6; Middle Duffryn 1850; Lower Duffryn and Cwm Pennar 1850–54; Abergwawr 1855; and Cwmdare 1852–4.

Thomas Powell was the leader in the South Wales coalfield. By sixty he was severely deaf, but he had achieved superiority in both the domestic and steam coal markets in Wales at a time when other traders were likely not to own more than two mines.

Powell also took part in banking, docks and railways. He was a member of the initial board of the Taff Vale Railway and used his authority to support his coal business. The year before his death, in 1862 he possessed sixteen mines exporting 700,000 tons of coal a year. He was an adept and perhaps unscrupulous businessman, warning the mine engineer John Nixon, who wanted his designated commission for work he had done for Powell, 'I had a law suit with Lord Bute, and I beat him … I never in my life made an agreement that I could not get out of.' It is recorded, however, that Powell lost a case against Sir Charles Morgan from Tredegar, as he mined under his land without permission.

Powell did have a soft side. He was responsible for funding five widows who had lost their husbands in an accident at Duffryn in 1845 which killed twenty-eight people. Powell refused a public subscription, preferring instead to give the grieving women 9s a week plus their house and fuel. He married three times, with three sons and two daughters. Unfortunately, his sons were said to have been unlucky; the eldest was killed in Abyssinia with his wife and children in 1869, the youngest vanished in 1881 in a hot-air balloon over the English Channel, and the middle son suffered a fatal riding accident. After his death, George Elliott would play a role in his legacy.

The overseas market for Welsh coal was increasing, especially after 1834 when export duties were practically eliminated for coal transported by British vessels. Also, in that year, the tax charged for coal conveyed on foreign ships decreased to 4d for every ton. In 1837, French import tariffs on coal sent to ports between Loire and Seine were cut back. Robert Peel, the British Prime Minister, in 1842 brought back the export levy of 2s per ton on coal transported by British vessels. This was done to balance out his widespread concessions in tax, but the fierce backlash showed how much the Welsh collieries depended on the foreign market. Gatherings of dissent were held in Swansea, which was still the predominant nucleus for the export market, and communication was sent to the Prime Minister urging him to row back the tax. The Prime Minister took note and by 1845 this levy was dropped, leaving the sole outstanding export levy to be that of 4s for each ton of coal transported in foreign ships that had no corresponding arrangement with the UK. By 1850, this was ditched as well.

There was a huge expansion in the French market for coal from South Wales, with John Nixon paving the way. Steam coal was not in great demand in France before 1840, and it was only anthracite that was being exported in small amounts. Nixon discovered smokeless coal while travelling on a London steamer in 1840. He was amazed by the lack of fumes and was even allowed to stoke the fire to check the quality of the coal. It had come from Merthyr in Wales, and was traded by the London dealer Wood. Wood principally supplied the Thames steamers, conveying 150 tons of coal from Cardiff each week. This inspired Nixon, who went to Wales and talked to Thomas Powell, who was then excavating the Old Duffryn mine. He explains what happened in a speech recorded in the *Merthyr Guardian* for 12 May 1860:

> I went to Mr Powell, and told him that if he wanted a market for his coal in Duffryn, I was willing to enter into some arrangement with him so as to introduce his coal into France, and that I was engaged in the trade of that country. After a great deal of negotiation we agreed that he should pay me so much per ton on all the coal exported to France, that he should at first give the coal, and pay the freight, and that I should go over there and give it away. I knew that was the only way to introduce it into that country.

Thomas Powell gave Nixon some coal with the premise that he would show the French how to handle it. Nixon chose the sugar refiners to start with. They needed consistent steam pressure to work their boilers, and they favoured Welsh coal over Newcastle coal as it was cheaper to run and created less work in disinfecting the apparatus. Proprietors of river steamers were the next target for Nixon. They tested the Welsh coal and Nixon secured a deal to provide them with 3,000 tons at 2s 6d more than Newcastle coal. Nixon was also able to obtain a contract from the French government as an alternative to using Newcastle coal. By 1851, 452,305 tons of coal were exported from Wales compared to the 68,829 tons exported in 1840.

The UK was still the predominant market, however, buying 1,573,774 tons of Welsh coal in 1851 compared to 1,374,264 tons in 1840. Cardiff had developed considerably, producing 501,063 tons of coal for the UK market in 1851 compared to 162,292 tons in 1840. Newport was still a big supplier too, yielding 451,475 tons in 1851, but it had not shown the growth of Cardiff; indeed, it had produced more in 1840.

There was enormous competition between northern coal and Welsh coal. The idea of official testing was devised, although private trials of different coals were already common. The Blackwell Railway tested coals from the north with Welsh coal and found that 30 per cent less Welsh coal was needed for their operations compared with the northern coal. However, the outcomes of these tests were not widely published or trusted. When the Admiralty started testing coal from the mid-1840s, people started paying more attention as the results were published under the jurisdiction of Parliament. Rather than conduct independent trials, steamship owners and other companies, both at home and abroad, were assisted in making their purchasing decision by what were called the Admiralty Lists. Coals that found a place on these lists sold in much greater quantities.

In 1845 the Admiralty had just purchased its first iron steamship, the *Birkenhead*, and Joseph Hume MP asked the Lords to test the coal being used on it, warning that they might be paying an inflated price for poor coal and in doing so letting down the nation:

> Without an accurate knowledge of the power of coals to be used, the country may be paying the highest price for an inferior article, and depending on the power of the fuel the public service may suffer disappointment at a moment when the greatest interests of the country may be at stake.

The Admiralty subsequently paid for tests which were conducted by Sir Henry de la Bèche and Dr Lyon Playfair. Ninety-eight types of coal were tested, with thirty-seven coming from Wales (which had a greater range of coals), and seventeen from Newcastle. Welsh coals were the clear winner as they ignited well, had good evaporative power and created a clean fire with little smoke. Coals from Newcastle caked easily, stopping airflow, and needed a lot of stoking while emitting thick black fumes. The Welsh coal did tend to create a lot of ash, but this was overlooked in view of its other qualities. The Admiralty therefore favoured Welsh coal over Newcastle coal. Indeed, during the Crimean War the Rhymney Railway Bill was bought to the attention of Parliament and Sir James Graham, First Lord of the Admiralty, said that Cardiff steam coal was so important to the navy that the railway was a matter of national significance.

Those involved in northern coal extraction, however, were furious at the result of the Bèche-Playfair trials and spent the next two decades lobbying Parliament, using the press and new trials to further their cause. They realised they had to find a way to reduce the smoke that their coals created. A Mr Williams of Liverpool devised a method of reducing the smoke in seaborne engines, and as a result of trials that were performed with naval officers watching, some maritime business was returned to the north.

The Welsh responded by proclaiming a test whereby two trans-Atlantic steamers would travel between Liverpool and New York, one using Welsh coal and the other northern. Colliery owners from the north enjoyed this trial as it resulted in naval officers formally announcing in October 1858 that northern coal was appropriate for use by the navy. Thomas Powell and John Nixon disputed this, writing a powerful retort in *The Times* indicating that the tests had been non-viable as 'the boilers and furnaces were peculiarly adapted for the consumption without smoke of the North Country coal, but of a construction not similar to any now in use in the Royal or mercantile steam navies'.

Powell believed Welsh coal to be superior. The Welsh colliery owners performed new tests in Cardiff, in which, predictably, Welsh coals came out as the winner. Authorised reports on these tests came out in March 1859, with no judgement given to each region. Welsh coal was shown to have a better evaporation rate, while the Newcastle coal burnt more quickly. The Welsh were not happy with these results, even though they were providing the Admiralty with 188,500 tons out of 249,547 tons in the year of 1859.

A compromise was reached by the Admiralty in 1863 after dockyard tests in 1860 in which Welsh coal came out on top. Newcastle coal owners accepted their defeat but asked for a combination of both coals to be trialled. They brought to notice the crumbly nature of Welsh coal, which led to the formation of large amounts of small coal when travelling. Boilers could be adapted to better utilise the northern coal so that smoke would be reduced. By March 1864, a formal announcement declared that an equal mix of Welsh and northern coal would be used by the navy. The Newcastle colliery owners were delighted. They had the document interpreted into various languages and sent the report all over the world.

Fortune changed in favour of the Welsh coal in 1867, when its smokeless qualities allowed for the capture of a slaving boat in African waters, after which orders were issued that only Welsh coal should be used by the Admiralty. However, the north was not going to give up that easily, and in 1869 it was announced that one-third of the coal used by the navy should be from the north, apart from on ships stationed in China and Africa. New trials were undertaken. Officers talked about the smoke created by a mix of coal, which in conjunction with the guns used in battle created a fog that made viewing other ships and the signals that they were giving incredibly difficult. One officer, the commander-in-chief of the Mediterranean Squadron said,

> I consider it my duty to point out to their Lordships that, in case of war, North Country coal is totally unfit for Her Majesty's service.

As the years went on, it was shown that the dockyard trials did not give an accurate picture of how the coals would perform at sea. By

1872, the government decided that once the reserves of northern coal were used they would only provide Welsh coal to the navy. Up until that point, supplies of Welsh coal were to be made available if there were situations where a large quantity of smoke would be a drawback.

The north-east had been the main exporters of coal in 1841, with 72 per cent of all outgoing trade. Thirty years later, in 1870, South Wales had enlarged their share of the export market to 31.3 per cent, whilst the north-east had fallen to 49.6 per cent. Both areas had increased their output of coal to serve the steamship industry, railways and growth in population. South Wales benefitted enormously from the Admiralty contract as well as having good access to ports to serve the rest of the world. However, that advantage did not help in London as ships had to navigate their way around Land's End in Cornwall before getting to the capital. Wales increased their share in the London market to 587,631 tons in 1874, but the total market of London coal was 7.5 million tons that year. The South Wales Railway linking Chepstow to Swansea, built in 1850, helped somewhat, but the Severn Tunnel was not completed until 1886 and the Great Western Railway charged huge tariffs on coal.

The coal trade was incredibly lucrative. From the years 1849 to 1875, output in South Wales increased from 4.5 million tons to 16.5 million tons. Immigrants flooded to the area in search of work, and railways continued to develop to allow new regions to be mined. Harbours were built, and landowners were happy to grant mineral leases for their land in order to gain royalties from the coal bought to the surface. Although house coal and anthracite had a market, they were both seasonal commodities. House coal relied on the winter months and anthracite depended on the harvest of hops and its function in drying and malting them. By 1840, other markets for coal had taken over in significance. These were bituminous coal, needed by the iron and copper industries for heating and melting, and steam coal.

After 1850, the value of steam coal rose from 1852 to 1856, from 1863 to 1865 and between 1870 and 1873. The Crimean War was one of the reasons for an increased market in 1854 when the navy set aside a further £160,000 for coal. In 1855, steamships owned by the UK government were said to require 430,000 tons of coal, three times the usual amount for use overseas. At this time the price of Welsh coal reached an all-time high, but it fell again at the end of the Crimean War in 1856 and did not regain strength until 1863. There was also a peak in prices after the Franco-Prussian War of 1870 to 1871.

Worry that the Mines Regulation Act of 1872 would limit production and raise costs kept prices high. In 1872 a ton of steam coal bought in Cardiff would cost at least 25 shillings, but by the end of 1875 even a ton of large steam coal in Cardiff was fetching no more than 11 to 13 shillings.

Due to market forces, the price of coal could vary considerably. George Elliot himself commented in 1873, 'If there is a scarcity of coal you are on your beam ends … you must have it at any price.' This was in the boom years, and in stark contrast to when there was a period of low demand. George wrote that it was 'as difficult to keep down the price in the year 1872 as it was to get it up in the years 1869 and 1870'. The weather also had something to do with it, with a newspaper reporting in September 1872:

I doubt we have ever had such cold weather before on this period of the year. During the night the thermometer fell to five degrees below freezing: this on September 21. The equinox is really extraordinary. The day has been miserably cold and wet; but, as the wind has veered to the south-west, I suppose the temperature will become more reasonable.

Another newspaper article in that year directly remarks on George's activities:

The clerk of the weather is in conspiracy with the coal-owners to keep up the price of coal. He shewed his malignant spirit in the spring by sending us frosts in May just as the prices were beginning to rise, and thereby accumulating such a superfluous stock of caloric that he nearly killed the whole of New York in getting rid of it. And now, almost before we have passed out of summer, without having even the decency to wait for the equinox, and until the sun has crossed the equator, he began sending us true winter weather – sharp frosts every night, snow in Yorkshire and Scotland, and hail in other parts of the country. He does this in order to check the fall in the price of coal which began last week. An extra three weeks of household fires throughout the kingdom will effectually keep prices up. I should like to know what commission the aforesaid clerk gets upon this little transaction.

I have spoken about the price of coal. Let me return to that subject to mention that Mr. George Elliot, M.P., has just made a very satisfactory contract. Mr. Elliot, who has risen from the ranks to be probably the largest coal owner in the kingdom, owns large collieries, not only in the north of England, but also in South Wales, where a few years ago he gave £380,000 for the Powell Duffryn steam-coal collieries. Recently he has arranged with the P. and O. Company to supply them with 70,000 tons of steam coal at 18s. per ton. This represents a sum of £153,000. He will pay to the miner, at the present rate, about £25,000 for cutting his coal, the royalty will cost him, at 6d. per ton, £4,250, and the railway expenses will foot up to £8,500. This amounts to £38,250. The interest on his capital invested in the mines, a 5 per cent., will represent £19,000. This brings the total up to £57,250, leaving Mr. Elliot the handsome profit of about £96,000 on the transaction. Doubtless he has many other expenses not reckoned in this calculation, but, on the other hand, the Peninsular and Oriental Company is not the only customer for whom he is at present working.

However, the work of the collier did not go unrecognised, especially in the wintery weather of 1872. The *Wolverhampton Chronicle* carried one piece at the time:

Gift from a Queen to a Collier
A correspondent vouches for the correctness of the following anecdote: – during the visit of her Majesty the Queen to Dunrobin Castle, the seat of the Duke of Sunderland, a few weeks ago, an incident occurred which has not yet been made public. The duke is having a shaft sunk on his estate for the purpose of proving some valuable mines, and there are engaged in the work several colliers from Shropshire, servants of the Lilleshall Company. The Queen being told of the mining operations expressed a desire to visit the spot, whither she was escorted by the duke. Whilst standing on the bank inspecting the work it commenced to rain. A few yards off one of the men named Cooper was sawing some timber for the shaft, and not being aware of the immediate presence of Royalty, heedless of the rain, continued his work without a coat. Presently he was surprised to feel a light touch, and on looking up, perceived the same duke, who laid a costly rug over his shoulders, at the same time exclaiming 'The Queen requested me to present you with her own rug; you may keep it and wear it.' The man's surprise, and the remainder of the story may be more easily imagined than described. The proof of this affair, to wit, the 'rug' in the possession of the poor man, will not be soon overlooked among the humble but loyal colliers of Shropshire.

Gradually, over time, iron companies in Wales also began to sell their coal, creating competition with the colliers. The Aberdare Iron Company started sales in 1858, and in the following decade it marketed 150,000 tons per annum. By 1866, when Cyfarthfa entered the market and Nant-y-glo and Blaina followed in 1867, the vast majority of Welsh ironmakers had also become coal merchants. In 1873, ironmasters used approximately 3 million tons of coal in their ironworks while mining at least 5 million tons. Rivalries with the coal owners were exacerbated by ironmasters employing the small coal for use in their works, whereas the coal owners had to discard it; furthermore, ironmasters could pay lower wages to their workers. The amount of coal made available from South Wales increased dramatically due to the ironmasters' entry into the market.

Cardiff had grown far more quickly than Newport as a shipping destination for coal because of the substantial demand for its high-quality steam coal. Better rail links helped as well. The Taff Vale Railway, opened in 1841, was vital in taking coal to Cardiff and was designed for that particular purpose, with most of its revenue coming from coal. By the 1850s, this railway had been extended to service the coalfields of the Aberdare and Rhondda valleys and Cardiff became the most important trading port in Wales. Newport, by comparison, was only joined to the mines by canals and trams in the 1840s. It was a catch-22 situation, with businessmen unwilling to open new mines in an area where transport was poor but nobody prepared to put money into building a railway.

In 1839, the Marquis of Bute invested £300,000 in the new Bute West Dock in Cardiff. He expected to realise his return by the escalation in importance of his 20,000 acres of mineral-rich land in the area rather than from profits at the dock. However, the River Usk in Newport was also deep enough to allow sea vessels to load coal. The Newport Dock, completed in 1842, comprised just 4.5 acres compared to the 19.5 acres at Cardiff and demanded higher charges than those at Cardiff.

The valleys connected by train to Cardiff were the ones that produced the most coal, followed by mines that had transport links to Newport. Cardiff was shipping 3,780,000 tons of coal in 1874, whereas Newport was responsible for 1,066,000 tons. By 1870, the requirement for steam coal from the Rhondda Valley had become too much for the infrastructure of the Taff Vale Railway and Cardiff docks. When local capitalists' enthusiasm about the unfinished Alexandra Docks at Newport was flagging and the work of construction was at a standstill for want of funds, Sir George Elliot was induced to come forward, and from that time he evinced a great and practical interest in the undertaking. In 1875, the Alexandra northern dock was opened for business.

The dock was opened with great fanfare on Wednesday, 14 April 1875, with a newspaper reporting:

The Alexandra Docks at Newport, Monmouthshire, were opened yesterday with great ceremony. The project was first started about seven years ago by a Company, and after proceeding with the works at a considerable period it was found desirable to raise additional capital. Sir George Elliot, M.P. for North Durham, was induced on representations made to him to invest considerable capita in the undertaking, and he also consented to accept the position of vice-chairman, Lord Tredegar being chairman. The docks are constructed on property purchased from his Lordship, and are more than 2,560 feet in length and 500 in breadth, the area of the fine sheet of water covering nearly 30 acres. The depth at the sill is thirty-five feet in spring, and twenty-five in neap tides. This will enable a merchant-man of the heaviest burden to enter the docks with perfect safety. The docks are furnished with four of the powerful hydraulic drops of Sir Wm. Armstrong, together with all the improvements that have recently been introduced. The foundations of three additional ones have been laid, and can be completed, ready for use, at a short notice when circumstances require it. It has been estimated by Mr Abernethy and Mr A. Bassett, the engineers, that this machinery, when fully developed, will be capable of loading ships with nearly two million tons of coal per annum. The length of the entrance lock to the dock is 350 feet, with a width of 65. This affords ample entrance for ships of large burthen. The lock is provided with four pairs of carefully constructed gates, one of iron and three of wood, all worked by hydraulic agency, similar to motive power being likewise applied on the sluices. The dock comprises two hundred acres of property suitable for building purposes. At nine o'clock yesterday morning an immense procession, comprising several thousands of the members of benefit societies and trades' unions accompanied the Mayor (Mr Benjamin Evans) and Corporation, with a large number of the leading inhabitants of the town, and also a long line of carriages, to the new docks. Eight bands of music were in the procession, which was the most imposing ever seen in South Wales. At the docks there was a detachment of the 103rd Regiment. Its magnificent band played popular airs on the pier head during the opening ceremony. In addition to an immense concourse of spectators from Newport, a large contingent arrived from all the principal towns in Monmouthshire and South Wales, the whole town having been decorated with flags, festoons of evergreens, mottoes, and devices. Venetian masts were placed at short intervals on each side of the main street extending two miles. A spacious pavilion was erected for the directors and officials of the Dock Company. Sir George Elliot was present, together with Sir Alexander Wood, Messrs J.C. Parkinson, G.W. Jones, E.W. Phillips, E. Underdown, J. McLean, J. Abernethy, A. Bassett, and a number of other promoters of the dock. It is a matter of regret, owing to serious indisposition, the chairman of the company (Lord Tredegar) was absent; and it was stated that his lordship was in a most critical condition. The Mayor and Corporation, together with leading merchants and other inhabitants attended to present a complimentary address to the directors. The address having been read, the Mayor presented it to Sir George Elliot. He assured them, as an old inhabitant long connected with commerce, that it afforded him no ordinary satisfaction to witness the opening of the new docks, which must tend so materially to enhance the great mineral resources of the neighbourhood, the Mayor concluded by saying the inhabitants of Newport looked forward with great hopefulness to the success of the dock. (Cheers).

Sir George Elliot, in replying, remarked that he was glad to receive this recognition of the services of Lord Tredegar, himself, and other directors. He regretted the absence of Lord Tredegar, which they all deplored, and paid high tribute to the memory of the late Mr McLean, who was the means of introducing this dock to his (Sir George's) notice. The dock had been a long and steady piece of business, but they always found that things which required time to complete came to a permanent issue.

The Directors and Corporation then approached the dock gates, and the ceremony of formally opening them was proceeded with, Mrs Evans, the wife of the Mayor, pulling a lever which communicated with the hydraulic machinery by which the gates worked, and they then rolled slowly back, amidst the booming of cannon and the cheers of the many thousands that thronged the banks of the docks. The George Elliot, which only recently arrived, was the first ship to enter the dock, closely followed by the brigantine Lord Tredegar and the steam tug Lady Tredegar.

The procession then reformed and marched back to the town to the Victoria Hall, where the directors of the dock had been invited to a public breakfast, which was prepared in a magnificent style. Amongst those who sat down were the Mayor of Newport, Sir George Elliot, M.P., Mr G.W. Elliot, M.P., Mr T. Cordes, M.P., Mr J.C. Parkinson (managing director), and a large company of other gentlemen. The hall was likewise graced with many of the leading ladies of the town and country.

The Mayor occupied the chair, and in proposing the success of the scheme in an appropriate speech, his worship referred to the rise and progress of the docks, alluding, in complimentary terms, to the energy and ability of the directorate. They had bought the greatest enterprise to a successful issue, and it afforded him great pleasure in seeing present Sir George Elliot.

Sir George, in replying to the toast, on behalf of the directorate, acknowledged the compliment paid them by the Mayor, and said it afforded him no ordinary satisfaction that the enterprise had that day been bought to such a successful issue. It had been long his impression that Newport was one of the most admirably situated ports in the Bristol Channel. This gave it no ordinary advantage in a commercial point of view. It was evident that the large increase in dock accommodation, materially augmented the commercial prospects of the district. The public at large would participate in the result of the enterprise. There was no question that the increased shipping facilities afforded by these docks would enable it to class one of the best of the Channel ports. In alluding to the unfortunate differences that had existed between capital and labour in Monmouthshire and South Wales during the last four months he strongly advocated arbitration, instead of resorting to the extreme measures which had unfortunately been adopted, and which had produced such serious consequences to the welfare of the district. He was of opinion that had an equitable arbitration been adopted, it would be rendered apparent that not only was it necessary for the employers to give notice to their men of a reduction of 10 per cent. as proposed, but the cause could have been shown for a reduction of 20 per cent. The hon. gentleman resumed his seat amidst applause.

A number of other toasts were proposed by leading gentlemen of the port, and duly responded to.

During the proceedings, a telegram was received from His Royal Highness the Prince of Wales, congratulating the Mayor, Corporation, directors of the dock company, and the inhabitants of Newport, on the successful opening of the docks, which had been named after his consort. The proceedings were curtailed to enable the ladies and gentlemen to witness the brilliant illumination and decorations that were profuse throughout all the leading streets of the town. Despite the large number of visitors, it is gratifying to find that no casualty occurred.

By 1875, the docks at Cardiff had enlarged to 97.5 acres, with most of this land owned by the Marquis of Bute. In the same year, the opening of the Alexandra Docks in Newport extended the harbour by 28.75 acres, allowing serious competition with Cardiff. Any incentive offered by one port would be swiftly taken up by the others. Cardiff had fantastic dock provisions and offered low docking charges. Ships chose ports that advertised a speedy-turn around as charges were due on ships that did not load their cargo in the stated time.

A further South Dock was added to the Alexandra Docks in 1892, with the original dock being renamed the North Dock. The Pontypridd, Caerphilly and Newport Railway was built to transport coal from the Aberdare and Rhondda valleys straight to the Alexandra Docks, competing with the Taff Vale Railway that served the two valleys. This line continued to transport passengers until 1962, and cargo until 1965. George Elliot took a great interest in the Pontypridd, Caerphilly and Newport Railway, which gave a straight route from his collieries to Alexandra Dock. Acting with Lord Tredegar, George had previously concluded that the only method of maintaining the profitability of the Alexandra Docks was by having a direct railway line there. A report in 1898 commented that the line was 'a costly hobby of the late Sir George Elliot, and now in the possession of the Alexandra (Newport) Docks Company'.

Coal owners themselves were an eclectic bunch. Money was needed to enter the trade, and with differing mines came differing backgrounds. Grocers, lawyers, shippers, merchants – all flocked to the coal trade. George Elliot liked people to know that he had started life as a pit laddie in the north. Thomas Powell was also happy to let it be known that he was a self-made man, saying,

'My friends were not born before me. I had nobody who ever gave me anything; I began the world with very little. I have been now 48 years in the coal trade, and I must have made very bad use of my time if I had not made a little.'

Thomas Powell was unique. He had achieved a monopoly on the bituminous coal market by 1840 with his mines in Monmouthshire and Gelligaer, and followed this by presiding over the steam coal trade in the Aberdare Valley when he was profoundly deaf and in his mid-sixties. He died in 1863, aged eighty-three, having spent the previous day at work in his Newport office.

He left his estate to his three sons. One, the younger Thomas Powell, was not interested in steam coal, instead concentrating on the Llantwit pits that produced house coal. His other two sons, Walter and Henry, did not want to take on such a big business. They had the steam collieries valued by three renowned mining engineers: William Armstrong, T. E. Forster and our own George Elliot. George was so impacted by the mines that he decided to form an undertaking to manage the collieries. He offered Walter and Henry Powell the sum of £365,000 for the sale of the collieries, registering a limited corporation on 28 July 1864, as explained by historians J. H. Morris and L. J. Williams:

> This company had been founded after a searching valuation made by some of the most eminent mining engineers of the day, one of whom – George Elliot – had bought the property on this basis; its operations were conducted by a man – George Elliot – of the highest repute in the industry; and it had sound capital structure backed by men of considerable substance.

In total, George and his team bought sixteen collieries, which employed 16,000 people. Thomas Powell, when he died, was the fourth-largest exporter of coal in the world.

George Elliot started with a nominal capital of £500,000. This was allocated into 100 shares, worth £5,000 each. Eight investors joined him: S. E. Bodlen, Thomas Brassey, P. G. Heyworth, J. R. McClean, Alexander Ogilvie, Richard Potter, J. Swift and

W. Wagstaffe. George Elliot took the role of general manager at the mines with a salary of £1,000 per annum. The men all knew each other well, either because they were already friends or through business associates. Thomas Brassey, for example, had advised Richard Potter to give gifts to the right person in the French administration to ensure payment for the timber that he had provided for them during the Crimean War. George Wilkinson, who had been a colliery manager for Thomas Powell, retained his position while David Griffiths was appointed as sales agent and H. W. Dallas became secretary.

These investors in the Powell Duffryn Steam Coal Company enlarged their assets by purchasing neighbouring regions. For example, in 1865 they paid David Williams £8,750 for the Ynyscynon and Treaman pits, which were adjacent to their own Abergwawr colliery. They purchased the Fforchaman and Cwmmneol mines near Aberdare in the same year for £80,000 from the United Merthyr Collieries Company, and in 1866 they bought the High Duffryn colliery. They paid £123,000 to Crawshay Bailey, the great ironmaster, in 1867 for the Aberaman estate comprising the ironworks at Aberaman and the mines at both Aberaman and Treaman. New shareholders were needed to buy Bailey's collieries and 1,000 shares were issued, worth £100, with a dividend of 12.5 per cent, leading to a total capital in the company of £600,000.

David Williams' collieries had not been bringing in money for some time. The Powell Duffryn Steam Coal Company bought them with the aim using them to help the drainage at other mines they owned in the vicinity. The Crawshay Bailey estate had been valued at £250,000 in the boom year of 1864, and the Powell Duffryn Company was keen to extract the steam coal that was in the region. The company did not want the ironworks and attempted to sell them, but they could not find a buyer and so shut down the works and used the site for engineering activities and as a place to mend the wagons and trains in their possession. The company was big enough to absorb the years of bad trade and in a position to buy these companies at a low price, with the assets to plan ahead rather than worry about the here and now.

Even so, the company nearly closed during the downturn of 1868–9, showing the inherent risks of the trade. The group claimed that there had been a mistake in the ledgers at the time of purchase, noting that the yearly cost of working the mine had been put at £11,000 less than what it was in reality. This impacted quite heavily on their budget, especially with the trade in dire straits and their partners having to obtain capital in difficult economic times. The company asked for the contract to be revoked, or to reduce the acquisition of the mines by £72,000. The late 1860s were a difficult trading time, and the company had to recoup their loss. In 1869 the vice chancellor decided it was best to invalidate the contract, but later the Court of Appeal in Chancery ruled that the contract should remain but that £52,020 should be paid to the company as a reduction of purchase price.

The Powell Duffryn Steam Coal Company can be remembered for its robustness and durability. In 1867, the Rhymney Railway Bill was being discussed by a parliamentary committee, and the secretary of the company reported that they were providing 800,000 tons of steam coal each year, transporting 600,000 tons of that from Cardiff. By 1875, the company were raising 1 million tons of coal a year. They were able to remove one intermediary by maintaining their own ships. They also employed an agent in the countries to which they shipped, allowing them to seek out new markets for their coal. Raising cash by limited liability had produced disappointments for many over the last two decades, but the Powell Duffryn Steam Coal Company was an enduring colossus. David Leslie Davies' *Dictionary of Welsh Biography* entry for George takes up the story:

> By buying the rich coalmine and ironworks of Crawshay Bailey in that neighbourhood, Powell Duffryn seized the nucleus of the old estate of the Matthews of Aberaman, a branch of the ancient family of Radyr and Llandaff, and gentry of the district before becoming extinct in 1788. There, in their mansion (which had been renovated extensively by a previous purchaser, Anthony Bacon), Elliot lived at

intervals; and there, after his day, Powell Duffryn made their headquarters. Powell Duffryn proceeded under the leadership of Elliot and his successors to secure more coalmines in the Aberdare Valley and other mines in the Rhymney Valley. The company also developed railways in the Aberdare and Rhymney Valleys to promote exports, and the coke, electricity and gas works. In 1920 P.D. gained ownership of the old Rhymney Iron Company and its extensive estate, and the company bought thousands of acres in the Llantrisant area. The company's foreign business was so vast by 1914 that a branch was established in Europe, Compagnie Française des Mines Powell Duffryn. This growth stemmed from Elliot's foresight and energy. He was acting manager of the company, 1864-77 and 1880-88; and chairman, 1886-89. Elliotstown, Rhymney Valley, was named after him, and streets in his and his wife's memory in Aberaman. In memory of his wife he paid for a new church there in 1882-3, and endowed a new church in Whitby, Durham, in 1886. However, he was not free from opposition. The trustees of the Marquis of Bute were reluctant to grant him everything he sought, so he took an interest in the development of Newport docks to avoid Cardiff, over which they had a hold. He was the chief promoter of the Alexandra northern dock in Newport which was opened in 1875 and which gave a foundation for the subsequent growth of the town; he obtained Parliamentary authority to lay the Pontypridd, Caerphilly and Newport Railway, 1878-8, to serve him in exporting coal. He was enthusiastic about the future of the coal industry to the last. Three months before his death he published a plan for a trust to hold all the resources of the industry in Britain, with the owners holding shares but sharing the profits with the workers and an insurance fund. Elliot was also a prominent public figure.

The coal owners at this time were mostly self-made men whose main interest was profit. They played a part in local affairs, but did not enter much into politics at Westminster unless it impacted on their financial concerns, such as when an export tax on coal was introduced in 1842. The other political issue which concerned them was securing Admiralty business. Colliery owners were more likely to enter local politics, where their extensive interests could be pursued, serving as mayors, councillors, aldermen, Poor Law guardians, magistrates and on local school boards (after 1870) as well as getting involved in local gas and water undertakings. As indicated by Morris and Williams, they were generous and gave money to religious institutions and educational or cultural objectives.

> It would be false, however, to look to calculated self-interest as their only motive in these matters. They were the natural leaders of the society in which they lived and regarded their participation as a public duty rather than as a private gain. They... belonged to the old school of colliery proprietors who came into daily contact with their men – a type unfortunately rapidly becoming extinct.

George took it one step further, becoming the Member of Parliament for North Durham in 1868. However, there was still a lot to do in Wales.

Over the years 1840 to 1875, output in South Wales increased three or four times. Whereas producing 50,000 tons a year of coal in a colliery in 1840 was uncommon, by 1875 over 75 per cent of collieries were producing at least 50,000 tons, with almost half of the collieries turning out over 100,000 tons yearly. Collieries in the north during this time were still larger than those in South Wales, but Wales was catching up. In 1874, there were still 183 owners in South Wales who held just one colliery. Out of the 339 collieries registered there at this time, 314 of them, or 79 per cent, were owned by coal owners who had no more than three mines in their possession. There was only one company in possession of over twelve collieries – Powell Duffryn.

Bigger collieries tended to mine steam coal. These pits were usually sunk after 1840, when mines were being dug on a greater scale. Steam coal was found lower than house coal, so capital expenditure needed to be recovered by greater workings. Steam coal was bought in larger quantities than house coal, often forming the basis of a ship's consignment, whereas house coal was frequently only traded by the cartload. Anthracite coal, like house coal, had a limited market.

It was capital that made the difference. Until the roaring trade in coal of the early 1870s, most investment had come from enterprising individuals. Enormous wealth was not required to create a mine. Men were able to open a medium-sized mine for less than £10,000. Shopkeepers, solicitors and other professionals with a small amount of capital set themselves up near the coalfield. The 1870s saw changes in the law, which allowed a structured capital trade, creating the emergence of limited companies on the stock exchange.

Banks followed the success of the mines, and credit was available to coal owners to establish new mines. Another way of raising money was to take on new partners in a colliery. Raising capital through shares was rare in the days before limited liability. Two northern powerhouses, the Durham Coal Company and the Northern Coal Mining Company, had failed terribly at the end of the 1830s, but Acts of Parliament in 1855 and 1856 created the practice of general limited liability. This suited the Welsh coalfield as it was leaving behind the practice of individual ownership or small partnerships of mines. Mines were becoming deeper, with more employees required and longer waits before the coal was reached. Thus, capital was needed, and limited liability meant that large amounts of capital could be gathered at a limited risk to shareholders.

There was a financial crisis in 1866 and interest in investment in the coalfields dwindled, not picking up again until the richly advantageous years of 1872–4. By 1874, without counting the ironworks, a quarter of the collieries in South Wales were limited companies. The Powell Duffryn Company, responsible for yielding more than 1 million tons of coal a year, was a vast producer; the normal output of the bigger firms was around 100,000 tons.

The chief trade of steam coal demanded rigorous quality. Small coal was not allowed. To begin with, quality was assured by hand picking the bigger pieces, which was an effective way of removing the smaller coal. As time went on, large screens were used like sieves to separate the smaller pieces. This coal was called 'colliery-screened large' as the screening was performed at the mine. However, the action was often repeated at the docks to qualify coal as being 'double screened'. Double-screened coal was more valuable, but this did not equate to profits for the coal owner. It was usual to pay a further 6*d* per ton for double-screened coal, but as coal had been filtered out during the screening the coal owner would see a loss of around 11 per cent of the coal or 10*d* per ton.

The coal business was very aggressive. Sometimes coal owners would co-operate and agree on a certain price, or limit output to increase prices, but such agreements were not strong enough to reduce competition. Prices were hard to restrict. Sellers regularly met at the ports to discuss fixing a price, hastening an already inevitable rise or fall in price. Restrictions on price had to be backed by collieries limiting production, but in Wales most coal owners were still enlarging their mines, and the practice of rationing would have been hard to organise. There was still a great number of separate coal owners, as well as unimpeded entry into the trade. The lives of the colliers were caught up in the atmosphere of stiff competition, with employers fighting any steps that might increase the costs of manufacture – including wages.

By the 1860s, things had settled down. There were a few strikes in the South Wales coalfield in 1864, but they did not last long. Trade in general was flourishing, and there were even wage cuts in 1868 that were accepted without strike action. The 1870s saw huge ups and downs in the price of coal, however, which caused serious issues for the workers. There were two other factors in the trouble at this time: attempts to destroy the first trade union created by the colliers, and the authority of the coal owners to reduce wages to align with those of the ironmasters, who were also powerful traders in coal. These factors led to widespread strikes for twelve weeks in 1871, for three months in 1873, and for five months in 1875.

These strikes were caused by the massive differences in the price of coal caused by great fluctuations in supply and demand. This was at a time before a sliding scale of wages was established, and for now wages were fixed to the price of coal. Miners found this difficult to tolerate; not only were their earnings faltering, but they did not have secure employment. Demand in the house coal market was subject to seasonal variation, and days were being missed due to winding, pumping or ventilating the workings. Poor weather, a lack of customers and delays while empty wagons returned from the docks all affected the flow of work. Years of bad trade were often marked not only by wage cuts but also by reduced hours.

Wages were a large part of the owner's costs, equating to between 40 and 75 per cent of production value. As a result, employers did not like to increase wages, although the men saw them as a bargaining tool and acted expediently to gain them. The situation was highlighted when a new seam of coal or a new colliery was opened and the cutting price, which was the benchmark for wages and profit, had yet to be determined. Conflict arose even at established mines where the cutting price had already been agreed. The width of the seam, the level of clod, the state of the floor and roof, the prevalence of small coal and other elements all affected the cutting price. The rates for 'dead work' such as setting cogs (roof supports), driving headings and ripping floor and roof spaces also had to be settled. Different geographical situations also played their part in setting the cutting price and could be a cause of later friction. If the seam encountered a fault, if it fragmented or split, or if it narrowed or started to incline, the men's pay changed, sometimes rewriting the entire cutting price altogether.

Changing mining techniques also caused some friction. Long-wall working, where a long wall is worked in a single slice, was taking over from the traditional pillar-and-stall method but the colliers were extremely unhappy about it. The introduction of the safety lamp met with the same reluctance – not because men were inclined to recklessness, but because it gave off a substandard light, which meant they produced less coal and were paid less. 'Billy Playfair' was a system similar to screening that separated large coal from small and assessed with greater accuracy the money due to each miner. Before this system, the owner had estimated the amount of small coal in a tram which was not to be paid for. The colliers did not like the new practice, complaining that the scales were not precise and the holes into which small coals fell were too big.

By far the greatest source of conflict, however, was the establishment of double shift working. This started in the north of England, where men would have to work two seven-hour shifts back to back. Hauliers and other workers, including the boys, would be required to work two hours after the first shift started and then work a twelve-hour shift in total. Owners realised that capital costs were a great deal higher when digging deeper mines and increasing production, with greater sinking costs using more expensive machinery. A bigger output would pay for these extra costs and the double shift seemed the ideal way of doing that.

In the late 1860s, the miners opposed these changes. They did not like the idea of having to share working spaces, as it would involve being dependent on other miners for their payment. This included the practice of undercutting the coal at the end of a shift so that the roof would collapse and the coal could be collected the following morning. The double shift pattern meant that another collier would be paid for this coal. The initial shift would start at 4 a.m. and the last shift would finish at 7 p.m. This meant that leisure hours would be greatly reduced. Women would have to cook and provide baths at different times of day, catering for both their lodgers and their family. The colliers were also deeply concerned about the effect that a double shift would have on the price of coal, as increased production could result in a lower wage.

A greater risk of explosions was also a concern. As the men demanded that their safety be paramount, employers grew increasingly exasperated. The industry was full of tension and strife on both sides. The coal owners did all they could to reduce the men's power. They brought in men from all over England and Wales at considerable expense in the event of a strike. However, these men were hounded and threatened by the miners so that they were frightened of going to work. Hostility easily passed from noisy derision into actual rough usage. Some of these outsiders stayed in the area after the strikes had finished, but it took a long time for them to be accepted by the strongly bonded mining communities.

The discharge note was another source of contention, as it had to be shown to future employers. This was a tool to maintain standards but could be used in strike time to prevent a dissenting miner getting any further work. Truck also operated in this fashion, with a discharge note denied to a miner if he owed the shop any money. John Owen, a solicitor for the colliers, said in 1843:

> So far as the trunk system is concerned, I have found it much easier to convict a poor man for going into a gentleman's preserve and stealing a pheasant, that it is to convict a gentleman for charging a poor man 11*d* for a pound of bacon that is only worth 6*d*.

It was not unheard of for miners to be forced enter the workhouse if they received a black mark on their discharge note. A newspaper in 1861 commenting on the strike in Aberdare said, 'Let the Company withhold this discharge, and the collier might just as well try to work his passage to the moon as attempt to obtain employment on these Welsh hills.'

Another method of controlling the workforce was law. It was a criminal act until 1875 for manual workers to break their contract. Although no written contracts were used in Wales, it was the practice for the owner and the colliery to expect one month's notice. If the collier failed to do this, he could go to prison for three months. Owners often took such cases to the courts only to relent and offer to re-employ the collier, frustrating the miners.

Magistrates and judges ruled against the colliers in almost every instance, and colliers realised that the only way they could gain leverage was through the formation of trade unions. There were several difficulties in this. One was the geography of Wales; its hills and valleys meant communities were isolated. Another was the various types of coal found across the region. Anthracite, steam and bituminous coal all had different markets and as a result did not face the same labour issues.

No real progress was made until the 1870s, and it was not until the 1880s that English unions began to pay attention to the miners in South Wales. The Leeds Conference in 1863 created the Miners' National Union but faltered, as one of its founders was found to be unreliable. George Elliot remarked in 1867, 'I do not think that they have any union except the spontaneous combination which arises if there is the least interference with them.'

The industrial disputes of 1871 and 1875 would severely stretch the limits of both the miners and the coal owners. These disputes were about the disparity in wages paid to colliers and those paid to ironworkers. It only ended when a sliding scale of wages was introduced to control wages.

Owners realised that they had to present a united front to the men. Ironmasters were used to a coordinated approach, having worked together since the beginning of the nineteenth century regulating prices and output. But wage modifications in the 1860s were harder to attain, mainly because the iron masters had become entrenched in the coal business. Coal owners met in 1864, creating the Aberdare Steam Collieries Association, in response to the men proposing trade unions and making demands in the boom years of 1863 and 1864.

This association was quite powerful. Its rights and commitments were legally set out in the Deed of Association and it had a close-knit geographical membership in the Aberdare Valley, made up of several large coal companies including the Powell Duffryn Company, Nixon's and David Davis, and its product was the world-renowned Welsh steam coal. But it did not have a wide membership: it only had eleven associates. In 1863, these mines produced 1.6 million tons of coal, which was nearly 15 per cent of the total yield of the industry. In 1870, the association included owners from the Rhondda valley, remodelling itself to create the South Wales Steam Collieries Association. But still, it was made up of just twelve businesses, with a total production of 2 million tons.

June 1866 saw a spike in food prices, and the colliers wanted an increase in salary to match it. Employers were wary as the financial world was unsteady. The South Wales Bituminous Collieries Association was created, using the steam coal structure as a blueprint. These organisations were in place during the big strikes of the early 1870s. The value of coal was on the increase in 1871, but coal owners refused to increase wages, as the ironmasters had done. There was a twelve-week strike, which was a challenging trial for both men and owners. The matter was put to arbitration, and it was agreed that the men should receive half of what they wanted, which was a 2.5 per cent raise. The trade union really shone through. Membership had increased from 309 in the Aberdare valley in February 1871 to 6,000 that July. By 1873, more than 30,000 colliers in South Wales were part of a trade union.

This strike also showed the unity of the coal owners in the Aberdare Steam Collieries Association. None of them broke ranks to reach an agreement with the men. By giving the men a 2.5 per cent increase, the coal owners had slightly reduced the gap paid by the ironmasters, who gave a 5 per cent increase. But there were disparities too; although the association looked after 75 per cent of all production from the Aberdare and Rhondda valleys, men were still able to find employment at other pits. The association regulated strikes in a few pits, but with the increasing power of the union and strikes being more widespread, they found themselves overpowered.

The years from 1871 to 1875 were difficult for both owners and colliers. There was lengthy strike action and contention among miners and their bosses, leading to great losses on both sides. The result was the introduction of the sliding scale of wages, which varied according to the price of coal, and on 11 December 1875 an agreement was reached that effectively standardised wages for the men.

Colliers liked the sliding scale as it took the control away from their employers, ensuring that no wage fall would occur without good reason. It was still the case that salaries accounted for a large part in overall costs, but it gave a valid framework on which to base wage changes. The shared interests of both employers and colliers also became apparent through this sliding scale, as by looking at it in detail owners were able to establish a basic standard of living for their men. Both parties were satisfied with the sliding scale, especially after the distressing years that they had all suffered. H. H. Vivian said,

> Reason has asserted her sway and a very desirable scheme has been adopted by both employer and employed. We shall no longer be a byword; we shall not longer be pointed at; and no longer will it be said, 'Look how they quarrel amongst themselves in the South of Wales district.'

Elliot had his role to play in the wages and working hours of colliers, with his entry in the *Dictionary of Welsh Biography* stating:

> It was Elliot who was chiefly responsible for reducing the hours of work of an underground worker from twelve to nine hours a day, and he was an important intermediary between the masters and the workers during the great strike of 1871 in South Wales. In 1874 he maintained that he had devoted a great part of his life to the welfare of the working classes; yet, he did not wish to be seen as an MP for that class – 'since there were other interests to be represented.

In 1864, when George and others formed the Powell Duffryn Steam Coal Company Limited, foreign markets were only just beginning to recognise the importance of steam coal from South Wales. The company was highly successful, building on their exports, which in 1864 stood at 2,220,000 tons, and by 1913 had risen to 19,330,000 tons. The company was nationalised in 1947.

The late 1860s were not good years for the coal industry, but the Franco-Prussian War in the early 1870s generated much activity and wealth in the coalfield. These years were followed by a depressed market for iron, with great financial hardship in the industry in 1875, leading the Powell Duffryn Company to abandon their interests in the iron trade. At the same time, the coal trade was on a real downturn, leading to the Cwmdare, High Duffryn, Upper Duffryn and Abergwawr collieries, along with the Lower Duffryn's upper pit, to be shut down. Three of these collieries – the Abergwawr, Upper Duffryn and High Duffryn – never reopened.

One of the greatest purchasers of Powell Duffryn coal was the Admiralty. The Admiralty tests of 1876 had shown conclusively that coal from this company was the best to use on the navy's ships. Other nations' fleets also came on board and bought the coal, identifying its high calorific power and smokelessness as the finest on the market. The majority of passenger steamboats also bought Powell Duffryn steam coal. Powell Duffryn steam coal was available all over the world, at ports in the Mediterranean, South America and the East, and at specially built steamer refilling sites. The coal was also utilised in railway transport, with markets in Britain, Europe, India, Argentina, Brazil and Uruguay. The other main buyers of this coal were factories, again with markets all over the world.

Welsh coal was said to be the best coal in the world, especially during conflict. The Germans mined their own coal, and yet still Welsh coal was preferred in Hamburg and Bremen, as well as Rotterdam in the Netherlands. Welsh coal travelled the world to almost a thousand ports and was sold abroad for much higher prices than locally produced coal. When people talked of Welsh coal, of course, they really meant Powell Duffryn coal.

The Powell Duffryn Steam Coal Company Limited sold quite a variety of different coal. Among their listing were:

- *The Powell Duffryn Large Steam Coal – a smokeless variety from both the Aberdare and Rhymney Valleys, used in the British and Foreign Admiralty.*
- *Best Large Steam Coal – sold for motor wagons in England and Wales; a smokeless variety.*
- *Paris Nuts was from the Aberdare Valley – sold exclusively in France; a washed, smokeless coal, 1 ¼ to 2 ½ inches wide.*
- *Aberdare Washed Beans – smaller, at 5/8 to 1 ¼ inches in size. From the Aberdare Valley; primarily used in steam-raising; a washed, smokeless coal.*
- *Aberdare Washed Peas – similar to Aberdare Washed Beans, with a similar market, but smaller, at 3/8 to 5/8 inches.*
- *Aberdare Washed Grains – comparable to Washed Peas, but smaller, at 3/16 to 3/8 of an inch.*
- *Aberdare Washed Dust – very fine; smaller than 3/16 inch.*
- *Aberdare Unwashed Dust- the same as the Washed Dust, but unwashed; used in patent fuel manufacture.*
- *New Tredegar Washed Nuts – from collieries in the Rhymney Valley; used in steamship bunkers and for steam raising. A washed coal, semi-bituminous – ¾ to 2 ½ inches in size.*
- *New Tredegar Washed Peas – with same properties as New Tredegar Washed Nuts, but smaller, at 3/8 to ¾ inch.*
- *New Tredegar Washed Duff – used widely for patent fuel manufacture and comprised of the small coal – less than 3/8 inch, left after the New Tredegar Washed Nuts and Peas had been removed.*
- *New Tredegar Colliery Small Coal – not washed; coal taken from the Rhymney Valley Collieries but not treated in any way.*
- *Aberdare Colliery Small Coal – untreated; coal removed from the Aberdare Valley.*

- *White Rose Large Coal – from the Rhymney Valley; a red ash House Coal.*
- *White Rose Cobbles – a House Coal ranging in size from 1 to 4 inches.*
- *White Rose Small Coal was a small House coal.*
- *Foundry and Furnace Coke – created at Bargoed Coke Ovens from coal, and subsequently washed to reduce impurities. It sold locally, nationally and throughout the colonies of Great Britain and interested foreign states.*
- *Coke Screenings range in size from ½ to 2 ½ inches; extricated from the coke and used in greenhouses, bakeries and foundries, which all utilise their small size.*
- *Boulets – made in Rouen for use in France, in the home and industry.*

Certain byproducts were also sold by the Powell Duffryn Steam Coal Company, including sulphate of ammonia, pitch, creosote oil and sulphuric acid. The sulphate of ammonia had a good level of ammonia (roughly 25%) and was sold all over the world.

George was acting manager of PDSC until 1877, aged sixty-three. He had a great interest in the safety of his workers, being eager to try different and evolving technologies. These included safety lamps, shaft detaching hooks and novel ways for cutting coal using different machinery. He made significant additions to the means of mining coal concurrently in adjoining seams, and set up trials of new ventilating methods at his collieries. He co-founded the North of England Institute of Mining and Mechanical Engineers and was a council member for many years, becoming president in 1868–9. In November 1868 he gave his inaugural speech, suggesting the unification of different regional mining organisations to strengthen recognition of the profession. He also advocated an alliance with the Institution of Civil Engineers, again to increase the value of the mining world. He himself became a member of this institute in 1856.

George was an active member of various engineering institutions and, unusually among colliery owners, he encouraged the founding of the Mines Inspectorate in the 1850s. He was called upon as an expert witness in particular government inquiries and assisted the royal commission on the coal industry back in 1871 as well as the 1886 royal commission on accidents in mines.

Three months before George Elliot died in 1893, he published a document laying out a plan of action for a trust, utilising all the assets of the coalmining industry in Great Britain. He envisaged owners holding shares and distributing profits among the colliers, with the additional feature of an insurance fund. He wanted to improve the working conditions of the miners, and so proposed that a share of the profits to be paid into a fund for retired colliers. One obituary recalls:

Of the many designs which occupied Sir George Elliot's busy life, none was more remarkable than the scheme for the amalgamation of the entire coalfields of Great Britain in which he was profoundly interested. In September 1893, he proposed a gigantic coal trust, the working of which was largely based upon the principle of according to the workmen, a voice in the regulation of their wages, and of the selling-price of coal, and also a share in the profits. With this object, it was provided that – after 1d. per ton had been set aside from the gross earnings for a Workman's Insurance Fund, and a profit of 8.3 per cent had been made upon the capital – any surplus profits not exceeding an amount sufficient to pay a dividend of 3.75 per cent on the ordinary shares, should be divided in the proportion of two-thirds to the men and one-third to the owners.

Any further profits were to be divided in equal thirds amongst the owners, the workmen and the purchasers, in the latter case in the form of a discount. Committees in the several colliery districts, upon which the miners should be equally represented with the owners, were to be appointed to fix the scale of wages and the selling-price of coal of three independent referees who were not to be members of the trust. The formation of an insurance fund, by which an adequate provision for old age could be effected, particularly interested him, and not many days before his death he was heard to say that his most earnest wish in connection with the scheme was that he might be able to leave the whole body of the mining population better off than he had found them. His knowledge of engineering, combined with the practical interest he took in the most important mining developments of his time, helped to place him in the position of an authority upon all questions connected with the trade; but possibly he owed as much to a homely northern strain of fidelity to interests which had once engaged his serious attention as to any other single quality, it was this characteristic which in all probability caused the miners of Durham to return him-under the good-humoured nickname of 'Bonnie Geordie' – more than once to Parliament.

4

TRANSATLANTIC CABLE

Before it united with North Borneo, Sarawak and Singapore in 1963 to become Malaysia, Malaya was absorbed into the British Empire during the reign of King George IV in the 1820s. The Malayan forests were 'a place much infested by tigers, to which it is necessary to proceed on foot'. Dr William Montgomerie, who was on the search for a specific tree there, described it as 'a venture of some risk to proceed to this spot; but I have offered a reward for specimens of the flowers and fruit of the tree, and am in hope of being able to produce some ere long'.

The tree in question, the Palaquium, also yielded a gum known locally as gutta percha or gutta taban. This tree can stretch up to 30 metres in height, with trunk diameters reaching a metre. The tree is evergreen, with small clusters of white flowers and in many species, edible fruit. Montgomerie had been assistant surgeon to the President of Singapore when Malaya fell into English hands. In his work he had discovered that gutta percha had similar properties to rubber, which was utilised in the production of surgical instruments. He collected some samples and made it available to his colleagues in the West. The gum also had certain electrical and mechanical attributes that made it especially useful for submarine telegraphy; indeed, its superior qualities would see its continued use in this field for more than a century. It is still used by dentists in root canal filling to this day.

Dr Montgomerie immediately realised the potential of gutta percha. In 1843, he wrote an accurate report on its properties and sent it, with specimens, to his sister's husband Henry Gouger, who took it to the Society of Arts with great excitement. The society's members saw the potential of the material in making moulded commodities and set about investigating its properties. The Victorians loved gutta percha. They used it to fashion ear-trumpets and tobacco preservers, reading in their newspapers the romantic reports of native Malayans climbing these huge trees with green-and-gold foliage, hacking away at the trunk to reveal the gum. They created mourning rings with it, embedding locks of hair from their deceased loved ones in the resin.

Dr Montgomerie had already been recognised for his presentation of nutmeg to the Society of Arts. He returned to England in 1844 to collect his gold medal for this, and brought with him several more examples of gutta percha. He gave some to the agent of Charles Mackintosh, who made waterproof jackets. Thomas Hancock, as Mackintosh's partner, was introduced to these samples when he visited later that week. Hancock passed on the news to his brother, Charles, who was in the process of inventing bottle stoppers made of cork. Charles subsequently added the gum to the patent. A Dutch chemist, Henry Bewley, got in touch with Charles in the following months, having been informed by the patent office. He was in the business of trying to keep soda water in the bottle. They met in February 1845 and reworded the patent to suit them both. The Gutta Percha Company was born.

Discussions at the Society of Arts continued about the exclusive attributes of the gum. Michael Faraday noted that its resistance to water could make it useful in insulating electric wires. Another guest at the conferences, Christopher Nickels, took some resin to his factories in Lambeth and developed ideas which corresponded with what Charles Hancock was doing. As a result, they decided to join forces, sharing patents and expenses.

Charles Hancock needed a regular source of income to help him develop his work, so Henry Bewley and Christopher Nickels decided to install him in a factory, paying him £800 a year as well as royalties. Henry Bewley took on one business partner after another, however, which irritated Charles. Arguments grew over the cork patent which he held, furthered by a patent held by Bewley which Charles had developed. This was an important invention as it devised a method of creating gutta percha tubes. Charles had taken this one step further, and created a machine which would use this tubing to cover the wires. Charles, to the annoyance of Bewley, had patented the machine in his own name, even though it was created in the factory during working hours.

By 1848, tensions were rising. Charles Hancock wanted control of the machine and Bewley responded by firing him with a year's salary. Charles created a new company with his brother Walter, with premises in West Ham, in direct competition with Bewley. The resulting animosity and price fixing nearly brought an end to both companies. The West Ham company went into liquidation several years later, with Charles taking on the remaining stock. He did his best to recover his position, claiming rights over his invention, but eventually the courts dismissed his case and the conflict over patents was finished.

Henry Bewley and the Gutta Percha Company carried on, producing goods such as tubing, insulating apparatus, acid-tank linings, suction belts and domestic implements as well as an array of ornamental goods. They exhibited at the Great Exhibition of 1851, firing the public imagination with items such as the Railway Conversion Tube, which allowed passengers to 'converse *with ease and pleasure*, whilst travelling, notwithstanding the noise of the train. This can be done in so soft a *whisper* as not to be overheard even by a fellow traveller. They are portable and will coil up so as to be placed inside the hat.' These tubes were already in use in 1850, when the Revd Samuel Meyrick informed the Company:

> The Gutta Percha Hearing Apparatus fitted up in Lismore Cathedral, for the use of His Grace, the Duke of Devonshire, has most fully answered the purpose for which it was

required. The tubes are conveyed from both the pulpit and the reading-desk to His Grace's pew (under the flagging and flooring, and altogether out of sight), and although their length is between thirty and forty feet, he is able, with their assistance, to hear distinctly every word.

The Gutta Percha Company also gained good business in selling the gum to make shoe soles, with clear evidence of the material's advantages over leather. Ice skates were also devised from the material, as well as the ingenious invention of snowshoes for horses.

Once the resin arrived from overseas, it was taken to the slicing machine in the factory where it was cut and then sent to the boiling room to be softened in heated water. The rolling room was the place where it was kneaded into sheets. A building was opened at 98 New Bond Street to showcase the full range of decorative uses for this new novelty, this 'milky juice'.

Happily, the unique properties of gutta percha led to the development of submarine cables for electric telegraphy. The year 1846 marked the inaugural year of the Electric Telegraph Company, which laid lines between Birmingham, Liverpool and Manchester. These initial lines were mainly uncovered; gutta percha was only used where they were supported by poles, but it was obvious that some lines would need to be underground and therefore would need the water-resistant properties of gutta percha.

The problem lay in finding the right combination of gutta percha, copper, hessian and lead to create a wire that would withstand wet conditions. Charles Hancock used extrusion, employing the screw-press apparatus designed by Bewley to produce rods and tubes. Landlines using this method were in common use, but attention was increasingly drawn to the potential manufacture of submarine cables.

Germany took the lead when Dr Werner von Siemens developed a machine rather like a pasta maker, in which he waterproofed the wire with gutta percha and successfully used it in the sea. Another proposal suggested lead piping to enclose the rubber and wire, which was accomplished in New York Harbour in 1842, leading to good results until ice destroyed the lead. Inventors tried to replicate this success in Portsmouth Harbour but were frustrated. It was not until 1849, when the electrician C. V. Walker put down 2 miles of piping in the English Channel, that messages were received and sent between the ship *Princess Clementine* and a receiving station in London.

A year later, in 1850, Jacob and John Watkins Brett requested 'twenty-five nautical miles of No. 14 Birmingham gauge copper wire covered with great care in gutta percha to half-an-inch diameter' to run between England and France, with the consent of both national administrations.

The Gutta Percha Company was only able to provide short lengths of wire at that point. Joining the wires was done by extending scarves over the join and soldering them with crushed gum and a blowpipe, after which they would be smoothed down to the shape of the wire. But there were dangers in this method, as the blow torch would melt the gutta percha at each end of the cable. Willoughby Smith came up with an ingenious solution, described in his book *The Rise and Extension of Submarine Telegraphy*:

The gutta percha was removed for about two inches from either end of the wires, which were then cleaned with emery paper, crossed and twisted several times one over the other, bell hanger fashion, and the whole covered with soft solder applied with an ordinary soldering iron, common plastic resin being used as flux. A suitable slice of plastic gutta percha was then placed on either side of the joint, the whole pressed in a wooden mould, where it was kept under pressure until the gutta percha was quite hard. The joint thus made resembled, when removed from the mould, a magnified cigar, some two inches in diameter and nine in length; it tapered at either end to the size of the covered wire over which it lapped. Each coil of wire was immersed in water as soon as covered, and, if thought sufficiently insulated, passed. The electrical arrangements were not elaborate, and although the battery generally used looked formidable, it was totally unsuited to much of the work for which it was required.

The cable was hauled onto a cart and taken to the Thames, where a steam tug called *Goliah* took it to Dover. Large crowds gathered to speculate about the mission, taking specimens of the cable home with them to discuss it with their friends. One end of the cable was positioned at the Cap Gris-Nez lighthouse in France, where it was dug into the ground. On the Dover side, the end was placed in a horsebox in the back yard of Dover station.

Action commenced on the morning of 28 August 1850, when the join was made between the cable at Dover and the cable on *Goliah*. It was a slow and steady start – the cable was laid at 4 miles per hour. Weights were attached to the cable, some sixteen in every nautical mile (naut). This got quite dangerous in the middle of the Channel, as the weights were heavy and men often got hurt, resulting in *Goliah* having to stop whilst the cable was laid. The size of the weight depended on soundings made by the ship *Widgeon*, which travelled alongside *Goliah*, with crew writing messages on their blackboard about the depth of the water, which indicated the size of the weight to use.

Goliah reached the continent that evening, to the incredulity of the French. A message was contrived using a type printing receiver, but to their annoyance it was illegible. The same happened for the English at Dover; they could not understand what had been written by the French. There was fury and blame on both sides. The problem was caused by dispersion in the signal, rendering the messages unintelligible, but this was not yet understood. The cable was abandoned.

However, not all were prepared to give up. A certain engineer named Thomas Russell Crampton wanted to develop the idea, using a similar cable but conveying messages more slowly. He advanced half the required £30,000 to fund the new project. As described by Willoughby Smith, the Gutta Percha Company put together a wire that was

> made up of four copper conducting wires of No. 16 Birmingham wire gauge [BWG], covered separately with two layers of gutta percha to quarter-inch diameter. The four cores were twisted together, and the interstices wound about at right angles with similar strands of hemp. Helically wound over all this were ten galvanised iron wires, No. 1 BWG, so that it looked from the outside like a great hawser and weighed about seven tons to the mile.

Worked was stopped unexpectedly by the news that Newall and Company, based in Gateshead, were claiming an infringement on their patent. Eventually it was Newall that finished the wire, with some of it manufactured by Kuper and Company in Camberwell. The wire went through a series of difficult situations, including dock tenants refusing to let the cable pass through their premises onto the Thames and bad weather taking the cable-carrying ship off course. The heavy weight of the cable meant that it paid out too quickly, leaving the men short of cable a mile from the French coast. However, an extra length was requested and soon the French end reached its destination. To the joy of all concerned, the cable was a success. On 13 November 1851, the first messages across the English Channel were received and sent.

The Gutta Percha Company, based in Greenwich, and Newall, in Tyne and Wear, began to share their commercial enterprise. They were still in the early stages of their invention, with cables losing their power after several hours due to too little insulation. Too much, however, made the cables cumbersome to lay. A terrible fire broke out in the Gutta Percha Company's Wharf Road premises in June 1853, destroying much of the building and two boats in the nearby canal. Some £30,000 of damage was caused, with the total assets valued at £100,000.

This is where George Elliot comes into the story. That contributing manufacturer, Kuper and Company had been taken over by Richard Atwood Glass, who had partnered up with George Elliot under the new name Glass, Elliot and Company. They bought 700 nauts of core from the Gutta Percha Company in 1854, who also supplied Newall with 200 nauts in that year. Glass, Elliot and Company renovated their wire rope facilities in Morden Wharf, East Greenwich, to produce cables.

They recognised the importance of testing the cables for their insulating properties and dug pits at Morden Wharf, supported by timber, to distinguish defects. During this process, they found that their pits were affected when the water level in the Thames dropped at low tide, so they had to adapt their testing to times when there was a high tide.

This all occurred during the Crimean War. Glass, Elliot and Company were rather put out when Newall was asked to provide 400 nauts of cable in 1855 to aid the war effort. The core came from the Gutta Percha Company and there was a great deal of secrecy surrounding the deal, not least because they managed to complete the business in a very short space of time. It was boarded onto the SS *Elba* for transportation to the Black Sea. Glass, Elliot and Company were highly inquisitive about the cable and when bad weather in the North Sea meant that *Elba* needed to stop in the Thames to be mended, they went to great lengths to see if they could find out more. However, it was not until the cable was laid between Varna and Balaclava that the secret came out – only the ends were protected with wire; the rest of the cable was merely joined and secured with gutta percha. The cable was in use for a year and did much good in bringing the war to an early end.

Glass, Elliot and Company were keen to progress and agreed with W. T. Henley, a competitor in the business, to share premises that had been leased by Enderby's Hemp and Rope Works near their existing property in Greenwich. Luckily for all, after a short period of time, Henley relocated to North Woolwich and Glass, Elliot and Company were left with the beautiful Enderby House, as well as a garden and meadows. They were visited by the eminent William Thomson, later to become Lord Kelvin, who was Professor of Natural Philosophy at Glasgow University. He had recently joined the Board of the Atlantic Telegraph Company, which had just been created by the New York, Newfoundland and London Telegraph Company. The Atlantic Telegraph Company was the idea of the American magnate Cyrus Field, who upon retiring had devoted his time to the miraculous endeavour of joining the two continents by cable. The Englishmen John Watkins Brett and Charles Tilston Bright joined forces with him, and the company was formed in December 1856. It led George Howard, Earl of Carlisle and Lord Lieutenant of Ireland, to voice his well-known address:

> Why, gentlemen, what excuse would there be for misunderstanding? What justification could there be for war, when the disarming message, when the full explanation, when the genial and healing counsel may be wafted even across the mighty Atlantic, quicker than the sunbeam's path and the lightening's flash.

Underwater cables had only been in use for a few years, starting with the wire from Dover to Calais in 1851. A big problem was that French fishermen would pull up the wire from the sea floor to take its copper content. This copper was the means by which the electricity was conducted. If they could avoid theft, the outer iron coating had to fulfil two main functions: it had to be small and light enough for the laying ships to convey it across the ocean – 2,000 miles in the case of the Transatlantic Cable – and it had to be strong enough to cope with the rigours of the seabed.

The idea initially came from a British engineer, Mr Gisborne, who had spent an evening discussing the project with Cyrus Field in a New York hotel. During that time, and as a result of their discussions, Field wrote to a friend in the navy about the state of the seabed in the Atlantic. He also penned a letter to Samuel Morse, asking how feasible it would be to sustain a current along a 2,000-mile-long wire. The response was obviously enthusiastic, and it did not take Field long to start the company, consolidating Great Britain, the United States and Newfoundland, who were generous with territorial concessions and in providing ships for the proposed mission. But what Field really needed was capital, and 'as the confidence of the moneyed men in the United States did not induce them to supply it' he tried his luck in Great Britain. He found the British to be more forthcoming, and within a month he realised £350,000, with the United States providing just £88,000.

The contract for the 1858 Transatlantic Cable was divided equally between Glass, Elliot and Company of East Greenwich and Newall and Company of Birkenhead, but Newall felt that if he didn't have full control over the order he didn't want anyone else to produce it. The two companies were rivals, and tensions were present. The Atlantic Telegraph Company was led by Sir William Brown in his capacity as chairman. Cyrus Field became general manager, with Charles Bright taking the role of chief engineer and Dr Edward Orange Wildman Whitehouse acting as electrician. Whitehouse was a GP by profession but enjoyed scientific experiments in his free time. Directors included John Brett, C. M. (later Lord) Lampson and John (later Sir John) Pender. It was the beginning of a new era.

By 19 December 1856, Glass, Elliot and Company were in the business of creating the 2,500 nauts of interior cable, made up with seven copper wires, which added a weight of 107 pounds to every naut. This was then covered by three layers of gutta percha, giving it a thickness amounting to three-eighths of an inch, and a total mass of 260 pounds for each naut.

Frederick Richard Window spoke to the Institution of Civil Engineers, chaired by Isambard Kingdom Brunel, on 13 January 1857 to celebrate the work:

The rapidly-extending system of submarine and international communication has now assumed so much importance, and forms such a prominent feature in the daily transactions of life, that a short review of the subject, in an engineering point of view, must be interesting to the Institution; although the Author regrets, that no one abler than himself has undertaken the task. This invention, for connecting lands which Nature herself has separated with her most formidable barrier, and for linking together remote kingdoms in the common chain of civilization, is a very modern discovery: yet so rapidly has it spread throughout Europe, that, from constant use, and familiarity with its manifold advantages, it has become, within a few years, a material fact, indispensable for our convenience and happiness, and is now an absolute necessity. Like the Electric Telegraph itself, its parentage is difficult to trace: It was the emanation of no single brain, but had been a cherished dream, from an early date in the era of electric telegraphy, the probable realization of which everyone admitted, although the means for its accomplishment were still unknown.

This wire was evaluated in July that year, with a total of 2,717 nauts under scrutiny by Whitehouse. He judged the insulation by using quite a large battery to pass electricity through the wires, with a fairly delicate galvanometer to read the results. Using this method, he was able to choose which wires to use. He did not, however, make any allowances for temperature, the age of the interior, or the time at which he conducted his experiments.

The first Transatlantic Cable expedition, held in the summer of 1857, was a project of immense collaboration. How to proceed with the initiative was a topic of great concern, as the distance to be covered was twenty times greater than any other cable-laying expedition. Both Newall and Glass, Elliot and Company manufactured 1,250 nauts of cable each. The wire core was to be made of seven strands of No. 22 BWG with six of these copper wires containing the seventh. On top of that, three coverings of gutta percha were applied, which was then covered in a jute thread which had been thoroughly immersed in a mixture of Stockholm tar, boiled linseed oil, beeswax and pitch. This core was then covered iron wires to the standard of No. 22 BWG and held in place by a further mixture of tar, pitch, beeswax and linseed oil.

The Newall cable was loaded onto the American ship USS *Niagara*, while the Glass, Elliot and Company portion went on board HMS *Agamemnon*. The British ship was borrowed from the Admiralty, and was both steam and sail, whereas the American vessel was a steam frigate on loan from the government. They were accompanied by HMS *Leopard* and HMS *Cyclops*, the latter being used to take soundings. USS *Niagara* was further aided by USS *Susquehanna*.

A newspaper article tells of the celebrations at Valentia Bay, on the Irish coast, as they set out for the first cable laying expedition in 1857:

Valentia Bay was studded with innumerable small craft decked with the gayest bunting. Small boats flitted hither and thither, their occupants cheering enthusiastically as the work successfully progressed. The cable boats were managed by the sailors of the *Niagara* and the *Susquehanna*. It was a well-designed compliment, and indicative of the future fraternisation of the nations, that the shore rope was arranged to be presented at this side of the Atlantic to the representative of the Queen by the officers and men of the United States Navy, and that at the other side the British officers and sailors should make a similar presentation to the President of the great Republic.

However well intentioned, this expedition was to end in failure in the Atlantic Ocean. Charles Bright, engineer-in-chief, reported to the directors of the Atlantic Telegraph Company:

In proceeding to the fore part of the ship I heard the machine stop. I immediately called out to relieve the brakes, but when

I reached the spot, the cable was broken. On examining the machine – which was otherwise in perfect order – I found that the brakes had not been released, and to this, or to the hand wheel of the brake being turned the wrong way, may be attributed the stoppage and consequent fracture of the cable.

When the rate of the wheels grew slower as the ship dropped her stern in the swell, the brake should have been released. This had been done regularly whenever an unusually sudden descent of the ship temporarily withdrew the pressure from the cable in the sea. But owing to our entering the deep water the previous morning, and having all hands ready for any emergency that might occur there, the chief part of my staff had been compelled to give in at night through sheer exhaustion, and hence, being short-handed, I was obliged for the time to leave the machine without, as it proved, sufficient intelligence to control it.

I perceive that on the next occasion it will be needful, from the wearing and anxious nature of the work, to have three separate relays of staff, and to employ for attention to the brakes, a higher degree of mechanical skill,

The origin of the accident was, no doubt, the amount of retarding strain put upon the cable, but had the machine been properly manipulated at the time, it could not possibly have taken place.

It was a costly mistake, but the appetite for further attempts was undiminished. The unused cable was reported by John Kell and Henry Clifford to be in good condition, 'not injured at the present time by the salt water that had been thrown on the coils to keep them cool'. However, it was put into storage in Plymouth and lost condition, especially in its insulating powers.

The Atlantic Telegraph Company was swift to try again. On Thursday 10 June 1858, a week after reaching Plymouth to

load up, ships set out to rendezvous mid-ocean at coordinates latitude 53° 17′, longitude 33° 18′. The day they left Plymouth the sea was calm, as it was the following day. On the 12th the wind started to pick up, and so on HMS *Agamemnon* the screws were lifted out of the water, the fires raked out and under 'royals and studding sails' she sped along. The weather continued to worsen and by Sunday 20 June a full storm was blowing.

What follows is a passage from Charles Bright's first-hand account of the work on board the *Agamemnon*.

The *Niagara*, which had hitherto kept close whilst the other smaller vessels had dropped out of sight, began to give us a very wide berth, and, as darkness increased, it was a case of every one for themselves.

Our ship, the *Agamemnon*, rolling many degrees, was labouring so heavily that she looked like breaking up. The massive beams under her upper deck coil cracked and snapped with a noise resembling that of small artillery, almost drowning the hideous roar of the wind as it moaned and howled through the rigging. Those in the improvised cabins on the main deck had little sleep that night, for the upper deck planks above them were 'working themselves free,' as sailors say; and, beyond a doubt, they were infinitely more free than easy, for they groaned under the pressure of the coil, and availed themselves of the opportunity to let in a little light, with a good deal of water, at every roll. The sea, too, kept striking with dull heavy violence against the vessel's bows, forcing its way through hawse holes and ill closed ports with a heavy slush; and thence, hissing and winding aft, it roused the occupants of the cabins aforesaid to a knowledge that their floors were under water, and that the flotsam and jetsam noises they heard beneath were only caused by their outfit for the voyage taking a cruise of its own in some five or six inches of dirty bilge. Such was Sunday night, and such was a fair average of all the nights throughout the week, varying only from bad to worse. On Monday things became desperate.

The barometer was lower, and, as a matter of course, the wind and sea were infinitely higher than the day before. It was singular, but at 12 o'clock the sun pierced through the pall of clouds, and shone brilliantly for half an hour, and during that brief time it blew as it has not often blown before. So fierce was this gust, that its roar drowned every other sound, and it was almost impossible to give the watch the necessary orders for taking in the close reefed foresail. This gust passed, and the usual gale set in now blowing steadily from the south west and taking us more and more out of our course each minute. Every hour the storm got worse, till towards five in the afternoon when it raged with such a violence of wind and sea that matters really looked 'desperate' even for such a strong and large ship as the *Agamemnon*. The upper deck coil had strained her decks throughout; and, though this mass, in theory, was supposed to prevent her rolling so quickly and heavily as she would have done without it, yet still she heeled over to such an alarming extent that fears of the coil itself shifting again occupied every mind, and it was accordingly strengthened with additional shores bolted down to the deck. The space occupied by the main coil below had deprived the *Agamemnon* of several of her coal bunkers; and in order to make up for this deficiency, as well as to endeavour to counterbalance the immense mass which weighed her down by the head, a large quantity of coals had been stowed on the deck aft. On each side of her main deck were thirty-five tons, secured in a mass, while on the lower deck ninety tons were stowed away in the same manner. The precautions taken to secure these huge masses also required attention as the great ship surged from side to side. Everything, therefore, was made 'snug,' as sailors call it; though their efforts by no means resulted in the comfort which might have been expected from the term. The night passed over without any mischance beyond the smashing of all things incautiously left loose and capable of rolling, and one or two attempts which the *Agamemnon* made in the middle watch to turn bottom upwards. In other matters it was the mere ditto of Sunday night; except, perhaps, a little worse, and certainly much more wet below.

Tuesday, the gale continued with unabated force; though the barometer had risen 29.30, and there was sufficient sun to take a clear observation, which showed our distance from the rendezvous to be 563 miles. During this afternoon the *Niagara* joined company, and, the wind going more ahead, the *Agamemnon,* took to violent pitching, plunging steadily into the trough of the sea as if she meant to break her back and lay the Atlantic cable in a heap. This change in her motion strained and taxed every inch of timber near the coils to the very utmost. It was curious to see how they worked and bent as the *Agamemnon* went at everything, she met headfirst. One time she pitched so heavily as to break one of the mainbeams of the lower deck, which had to be shored with screwjacks forthwith. Saturday, June 19th, things looked a little better. The barometer seemed inclined to go up and the sea to go down; and for the first time that morning since the gale began some six days previous the decks could be walked with tolerable comfort and security. But, alas! Appearances are as deceitful in the Atlantic as elsewhere; and during a comparative calm that afternoon the glass fell lower, while a thin line of black haze to windward seemed to grow up until it covered the heavens with a sombre darkness, and warned us that the worst was yet to come. There was much heavy rain that evening, and then the wind began not violently, nor in gusts, but with a steadily increasing force. The sea was 'ready built to hand,' as sailors say; so, at first the storm did little more than urge on the ponderous masses of water with redoubled force and fill the air with the foam and spray it tore from their rugged crests. By and by, however, it grew more dangerous, and

Captain Preedy himself remained on deck throughout the middle watch.

At 4 am, sail was shortened to close reefed fore and main topsails and reefed foresail. This was a long and tedious job, for the wind so roared and howled, and the hiss of the boiling sea was so deafening, that words of command were useless; and the men aloft holding on with all their might to the yards as the ship rolled over and over almost to the water were quite incapable of struggling with the masses of wet canvas, that flapped and plunged as if men, yards and everything were going away together. The ship was almost as wet inside as out and so things wore on till 8 or 9 o'clock, everything getting adrift and being smashed, and everyone on board jamming themselves up in corners or holding on to beams to prevent their going adrift likewise. At 10 o'clock the good ship was rolling and labouring fearfully, with the sky getting darker, and both wind and sea increasing every minute. Half an hour later three or four gigantic waves were seen approaching the ship, coming slowly on through the mist, nearer and nearer, rolling on like hills of green water, with a crown of foam that seemed to double their height. The *Agamemnon* rose heavily to the first, and then went down quickly into the deep trough of the sea, falling over in the act, so as to nearly capsize on the port side. There was a fearful crashing as she lay over this way, for everything broke adrift, whether secured or not, and the uproar and confusion were terrific for a minute; then back she came again on the starboard beam in the same manner only quicker and deeper than before. Again, there was the same noise and crashing, and the officers in the wardroom, realising the danger, struggled to their feet and opened the door leading to the main deck. The scene, for an instant, defied description. Amid loud shouts and efforts to save themselves, a confused mass of sailors, boys, and marines with deck buckets, ropes, ladders, and everything that could get loose, and which had fallen back to the port side were being hurled again in a mass across the ship to starboard. Dimly, and for a moment, could this be seen; and then, with a tremendous crash, as the ship fell over still deeper, the coals stowed on the main deck broke loose, and, smashing everything before them, went over among the rest to leeward. The coal dust hid everything on the main deck in an instant; but the crashing could still be heard going on in all directions, as the lumps and sacks of coal, with stanchions, ladders, and mess tins, went leaping about the decks, pouring down the hatchways, and crashing through the glass skylights into the engine room below.

Matters now became most serious; for it was evident that two or three such lurches and the masts would go like reeds, while half crew might be maimed or killed below. Captain Preedy was already on the poop, with Lieutenant Gibson, and it was 'Hands, wear ship' at once; while Mr. Brown, the indefatigable chief engineer was ordered to get up steam immediately. The crew gained the deck with difficulty, and not till after a lapse of some minutes; for all the ladders had been broken away, the men were grimed with coal dust, and many bore still more serious marks upon their faces of how they had been knocked about below. There was great confusion at first, for the storm was fearful. The officers were quite inaudible; and a wild, dangerous sea, running mountains high, heeled the great ship backwards and forwards, so that the crew were unable to keep their feet for an instant, and in some cases were thrown right across the decks. Two marines went with a rush head foremost into the paying out machine, as if they meant to butt it over the side yet, strange to say, neither the men nor the machine suffered. What made matters worse, the ship's barge, though lashed down to the deck had partly broken loose; and dropping from side to side as the vessel lurched, it threatened to crush any who ventured to pass. The regular discipline of the ship, however, soon prevailed and the crew set to work to wear round the ship on the starboard tack, while Lieutenants Robinson and Murray went below to see after those who had been hurt. The marine sentry outside the wardroom door on the main deck had not had time to escape and was completely buried under the coals. Some time elapsed before he could be got out; for one of the beams had crushed his arm very badly, still lay across the mangled limb jamming it in such a manner that it was found impossible to remove it without risking the man's life. The timber indeed, to be sawn away before the poor fellow could be extricated. Another marine on the lower deck endeavoured to himself by catching hold of what seemed like a ledge in the planks but, unfortunately, it was only caused by the beams straining apart, and, of course, as the *Agamemnon* righted, they closed again and crushed his fingers flat. One of the assistant engineers was also buried among the coals on the lower deck and sustained some severe internal injuries. *The lurch of the ship was calculated at forty-five degrees each way four or five times in rapid succession.* The galley coppers were only half filled with soup; nevertheless, it nearly all poured out, and scalded some of those who were extended on the decks, holding on to anything in reach. These with a dislocation, were the chief casualties; but there were others of bruises and contusions, more or less severe, and a long list of escapes more marvellous than any injury. One poor fellow went head first from the main deck into the hold without being hurt; and one on the orlop deck was 'chevied' about for some ten minutes by three large casks of oil which had got adrift, and any one of which would have flattened him like a pancake had it overtaken him...'

'Going round upon the starboard tack had eased the ship to a certain extent. The crew, who had been at work since nearly four in the morning, were set to clear up the decks from the masses of coal that covered them. About six in the evening it was thought better to wear ship once more and stand by for the *rendezvous* under easy steam. Her head accordingly was put about and once more faced the storm. As she went round, she of course fell into the trough of the sea again, rolling so awfully as to break her waste steam pipe, filling her engine room with steam, and depriving her of the services of one boiler when it was sorely needed. The sun set upon as wild and wicked a night as ever taxed the courage and coolness of a sailor. There were, of course, men on board who were familiar with gales and storms in all parts of the world; and there were some who had witnessed the tremendous hurricane which swept the Black Sea on the memorable November 14th, when scores of vessels were lost and seamen perished by the thousand. But of all on board none had ever seen a fiercer or more dangerous sea than raged throughout that night and the following morning, tossing the good ship from side to side like a mere plaything among the waters. The night was thick and very dark, the low black clouds almost hemming [the] vessel in; now and then a, fiercer blast than usual drove the great masses slowly aside, and showed the moon, a dim, greasy blotch upon the sky, with the ocean, white as driven snow, seething like a cauldron. But these were only glimpses, alternated with darkness, through which the waves rushed upon the ship as though they must overwhelm it, and dealing it one staggering blow, went hissing and surging past into the darkness again. The grandeur of the scene was almost lost in its dangers and terrors, for of all the many forms in which death approaches man there is none so easy in fact, so terrific in appearance, as death by shipwreck.

... A little after ten o'clock on Monday the 21st the aspect of affairs was so alarming that Captain Preedy resolved at all risks to try wearing the ship round on the other tack. It was hard enough to make the words of command audible, but to execute them seemed almost impossible.... Once round on the starboard tack, and it was seen in an instant that the ship was in no degree relieved by the change. Another heavy sea struck her forward, sweeping clean over the fore part of the vessel, and carrying away the woodwork and platforms which had been placed there round the machinery for under running. This and a few more plunges were quite sufficient to settle the matter; and at last Captain Preedy reluctantly succumbed to a storm he could neither conquer nor contend against. Full steam was got on, and, with a foresail and foretopsail to lift her head, the *Agamemnon* ran before the wind, rolling and tumbling over the huge waves at a tremendous pace. It was well for all that the wind gave this much way on her, or her stern would certainly have been stove in. As it was, a wave partly struck her on the starboard quarter, smashing the quarter galley and wardroom windows on that side; and sending such a sea into the wardroom itself as to wash two officers off a sofa. This was a kind of parting blow; for the glass began to rise, and the storm was evidently beginning to moderate; and although the sea still ran as high as ever, there was less broken water, and altogether, towards midday, affairs assumed a better and more cheerful aspect. The ward room that afternoon was a study for an artist; with its windows half darkened and smashed, the sea water still slushing about in odd corners, with everything that was capable of being broken strewn over the floor in pieces, and some fifteen or twenty officers, seated amid the ruins, holding on to the deck or table with one hand, while with the other they contended at a disadvantage with a tough meal, the first which most had eaten for twenty four hours. Little sleep had been indulged in, though much lolloping about. Those, however, who prepared themselves for a night's rest in their berths rather than at the ocean bottom, had great difficulty in finding their day garments of a morning. The boots especially went astray and got so hopelessly mixed that the man who could 'show up' with both pairs of his own was, indeed, a man to be congratulated.

But all things have an end; and this long gale of over a week's duration at last blew itself out, and the weary ocean rocked itself to rest. Throughout the whole of Monday the *Agamemnon* ran before the wind, which moderated so much that at 4 a.m. on Tuesday her head was once more put about; and for the second time she commenced beating up for the *rendezvous* then some 200 miles further from us than when the storm was at height on Sunday morning. So little was gained against this wind, that Friday the 25th, sixteen days after leaving Plymouth still found us some fifty miles from the rendezvous. It was, therefore, determined to get up steam and run down on it at once. As we approached the place of meeting the angry sea went down. The *Valorous* hove in sight at noon; in the afternoon the *Niagara* came in from the north; and at even, the *Gorgon* from the south and then, almost for the first time since starting, the squadron was reunited near the spot where the great work was to have commenced fifteen days previously as tranquil in the middle of the Atlantic as if in Plymouth Sound.

The end of the *Niagara*'s cable was sent on board the *Agamemnon*, the splice was made, a bent sixpence put in for luck, and at 2.50 Greenwich time it was slowly lowered over the side and disappeared for ever. The weather was cold and foggy, with a stiff breeze and dismal sort of sleet, and as there was no cheering or manifestation of enthusiasm of any kind, the whole ceremony had a most funereal effect, and seemed as solemn as if we were burying a marine, or some other mortuary task of the kind equally cheerful and enlivening. As it turned out, however, it was just as well that no display took place, as everyone would have looked uncommonly silly when the same operation came to be repeated, as it had to be, an hour or so afterwards. It is needless making a long story longer, so I may state at once that when each ship had paid out three miles or so, and they were getting well apart, the cable, which had been allowed to run too slack, broke on board the *Niagara*, owing to its overriding and getting off the pulley leading onto the machine.

The break was, of course, known instantly, both put about and returned, a fresh splice was made, and again lowered over

The work of drawers, as seen on this page, would have been familiar to young George as he languished in the mines working the doors. These illustrations are accurate depictions that accompanied the 1842 Report of the Children's Employment Commission. (Wellcome Collection)

The Davy lamp, a welcome innovation for miners who otherwise worked by candlelight, although it did not solve all their problems. (Tim Hall Willson)

The hellish environment of the coal mines in which George toiled would in fact have been much more cramped than this illustration. (Wellcome Collection)

'Inundation of a Pit', illustrating a disaster in which a Welsh colliery has been flooded in the mid-nineteenth century. Such tragedies were not uncommon at the time when George worked in the mines; indeed, he was responsible for avoiding some such tragedies by opening and closing the doors of the mine system.

A contemporary illustration of George in his prime as a businessman. (Author's collection; illustration from *Vanity Fair*, 29 November 1879, by 'Spy', aka Leslie Ward)

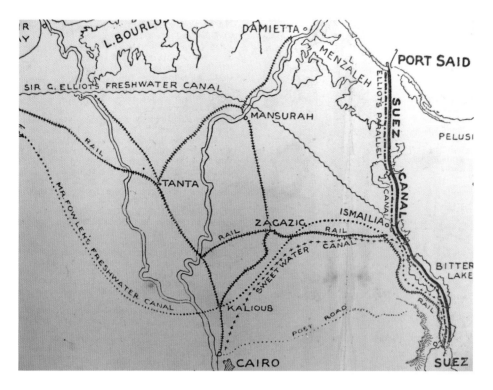

Maps from George's collection outlining the route of the Suez Canal. (Tony McCallum)

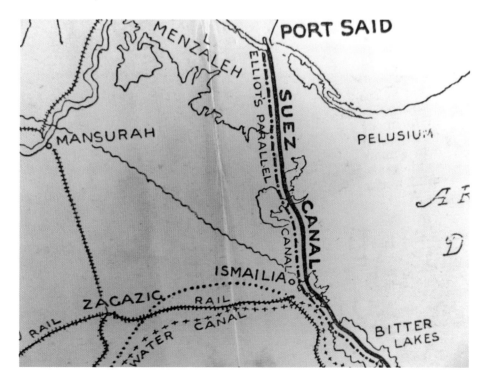

Above: Construction of the Suez Canal in progress at Lake Timsah. (Library of Congress)

Below: An evocative early photograph of the Suez Canal, a few decades after its completion. (Brooklyn Museum)

Right: A stall set up by the Gutta Percha Company at the Great Exhibition of 1851 showcasing their wares. (Rijksmuseum)

Below: Gutta percha being worked for George's Telegraph Construction and Maintenance Company at Greenwhich, painted 1865 by Robert Charles Dudley. (Metropolitan Museum of Art)

An 1863 photograph of USS *Niagara*, one of the cable-laying ships on the first attempt to lay the Transatlantic cable in 1857. (National Museum of the US Navy)

Robert Charles Dudley's painting of HMS *Agamemnon* laying the cable during the 1857 attempt on its way to meet USS *Niagara*. (Metropolitan Museum of Art)

Right: A cross section of Transatlantic cable. (Tim Hall Wilson)

Below: The *New York Herald*'s front page celebrating the successful second laying of the Transatlantic cable in 1865. (Library of Congress)

Bottom: The deck of the SS Great Eastern during the cable-laying journey, as painted by Robert Charles Dudley. (Metropolitan Museum of Art)

George's mother, Elizabeth.
(Tim Hall Willson)

George's wife, Margaret. (Tim
Hall Willson)

George in his later years. (Tim Hall Willson)

1. Park Street.
Park Lane, W.
19 [Aug] 1878

Dear Mr Disraeli

Allow me to offer my humble congrat-ulations. At the same time I am bound to add that the House of Commons has now ceased to command my enthusiasm. The only figure of all others that I cared most for is now removed from it. Henceforth my anxiety to be counted amongst its members will be immeasurably abated.

I conclude by always wishing you continued prosperity, health and long life & remain
Your most obt & great admirer
George Ellis

The Right Honble B. Disraeli

---

1. Park Street.
Park Lane, W.
29. Nov 1879

My Lord

I have received a communication from Mr Montague Corry to the effect that your Lordship has expressed a desire that on Monday next I should pay a visit to Hughenden. I have the honour to express my gratitude to your Lordship for such a distinctive token mark of friendship. I propose to leave by the 4.45 on Monday 1 Decr arriving at Six

I have the honour to be your Lordship's most humble servant
Geo Ellis

The Earl of Beaconsfield
Hughenden

This spread: Letters between George and Prime Minister Benjamin Disraeli, later Earl of Beaconsfield. (Weston Library, Oxford)

1. Park St
. Park Lane
29 Dec 1880

Dear Lord Beaconsfield

Allow me
to express my sincere
thanks for the very
kind letter of condolence
I have had the honour
to receive from your
Lordship — it has been
a great consolation
to me and my family
treasure it highly.

What can I say or
do but bow to the will
of that great being whose
wisdom and goodness
I dare not venture to
contest. Your Lordships
most ob[edient]

The Earl of Beaconsfield .   George Ellis

Royal Crescent in Whitby, where George lived for a time and where he cherished a friendship with Irish writer Bram Stoker. (Steve Daniels under Creative Commons)

George's favourite house, Houghton Hall. To his disappointment, he was only ever a tenant - the owners were not willing to sell. (Hans A. Rosbach under Creative Commons)

Right: St Mary's Church, West Rainton, built with funding bequeathed by George. (Robert Graham under Creative Commons)

Below: St Magaret of Antioch church in Aberaman, Wales, also funded by George. (Gareth James under Creative Commons)

Whitby Pavilion, built by George and opened in 1879. It continues to entertain visitors and locals to this day. (Gerald England under Creative Commons 2.0)

Above left: The author's father, Tony (left), and uncle Mike unveiling a plaque to George at Whitby Pavilion. (Tony McCallum)

Above right: A stained-glass window in Durham Cathedral commissioned by George after his wife's death and honouring Bernard Gilpin, known as the Apostle of the North. (Durham Cathedral) at half past seven. According to arrangement, 150 fathoms were veered out from each ship, and then all stood away on their course, at first at two miles an hour, and afterwards at four. Everything then well, the machine working beautifully, at thirty-two revolutions per minute, the screw at 26, the cable running out easily at five and five and a half miles an hour, the ship going four. The greatest strain upon the dynamometer was 2500 lb, and this was only for a few minutes, the average giving only 2000 lb, and 2100 lb. At midnight twenty-one nautical miles had been paid out, and the angle of the cable with the horizon had been reduced considerably. At about half past three forty miles had gone, and nothing could be more perfect and regular than the working of everything, when suddenly, at 3.40 am on Sunday, the 27th, Professor Thomson came on deck and reported a total break of continuity; that the cable, in fact, had parted, and, as was believed at the time, from the *Niagara*. The *Agamemnon* was instantly stopped and the brakes applied to the machinery, in order that the cable paid out might be severed from the mass in the hold, and so enable Professor Thomson to discover by electrical tests at what distance from the ship the fracture had taken place. Unfortunately, however, there was a strong breeze on at the time, with rather a heavy swell, which told severely upon the cable, and before any means could be taken to ease entirely the motion on the ship, it parted a few fathoms below the stern wheel, the dynamometer indicating a strain of nearly 4000 lb. In another instant a gun and a blue light warned the *Valorous* of what had happened and roused all on board the *Agamemnon to* a knowledge that the machinery was silent, and that the first part of the Atlantic cable had been laid and effectually lost.

The great length of cable on board both ships allowed a large margin for such mishaps as these, and the arrangement made before leaving England was that the splice might be renewed and the work recommenced till ship had lost 250 miles of wire, after which they were to discontinue their efforts and return to Queenstown. Accordingly, after the breakage on Sunday morning, the ships' heads were put about, and for the fourth time the *Agamemnon* again began the weary work of beating up against the wind for that everlasting rendezvous which we seemed destined to be always seeking. Apart from the regret with which all regarded the loss of the cable, there were other reasons for not wishing the cruise to be thus indefinitely prolonged, since there had been a break in the continuity of the fresh provisions; and for some days previously in the wardroom the *pièces de resistance* had been inflammatory-looking *morceaux,* salted to an astonishing pitch, and otherwise uneatable, for it was beef which had been kept three years beyond its warranty for soundness, and to which all were then reduced.

It was hard work beating up against the wind; so hard indeed, that it was not till the noon of Monday, the 28th, that we again met the *Niagara;* and while all were waiting with impatience for her explanation of how she broke the cable, she electrified everyone by running up the interrogatory 'How did the cable part?' This *was* astounding. As soon as the boats could be lowered, Mr Cyrus Field with the electricians from the *Niagara* came on board, and a comparison of logs showed the painful and mysterious fact that at the same second of time each vessel discovered that a total fracture had taken place at a distance of certainly not less than ten miles from each ship, as well as could be judged, at the bottom of the ocean. The logs on both sides were so clear as to the minute of time, and as to the electrical tests showing not merely leakage or defective insulations of the wire, but a total fracture, that there was no room left on which to rest a moment's doubt of the certainty of this most disheartening fact. That of all the many mishaps connected with the Atlantic telegraph, this was the worst and most disheartening, since it proved that after all that human skill and science can effect to lay the wire down with safety has been accomplished, there may be some fatal obstacles to success at the bottom of the ocean which can never be guarded against, for even the nature of the peril must always remain as secret and unknown as the depths in which it is to be encountered. Was the bottom covered with a soft coating of ooze, in which it had been said the cable might rest undisturbed for years as on a bed of down? Or were there, after all, sharp pointed rocks lying on that supposed plateau of Maury, Berryman and Dayman? These were the questions that some of those on board were asking.

But there was no use in further conjecture or in repining over what *had* already happened. Though the prospect of success appeared to be considerably impaired it was generally considered that there was but one course left, and that was to splice again and make another and what was fondly hoped would be a final attempt. Accordingly, no time was lost in making the third splice which was lowered over into 2000 fathoms water at seven o'clock by ship's time the same night. Before steaming away, as the *Agamemnon* was now getting very short of coal, and the two vessels had some 100 miles of cable between them, it was agreed that if the wire parted again before the ships had gone each 100 miles from the rendezvous they were to return and make another splice; and the *Agamemnon* was to sail back, the *Niagara* it was decided, was to wait eight days for her reappearance. If, on the other hand, the 100 miles had been exceeded, the ships were not to return, but each make the best of its way to Queenstown [Ireland]. With this understanding the ships again parted, and, with the wire dropping steadily between them, the *Niagara* and *Agamemnon steamed* away, and were soon lost in the cold, raw, fog, which had hung over the rendezvous ever since the operation had commenced.'

Warned by repeated failures, many of those on board scarcely dared hope for success … For those who do not derive any particular pleasure from the mere asking of questions, the harmonious music made by the paying out machine during its revolutions supplies the information...Then again, the electrical continuity after all the most important item was perfect, and the electricians reported that the signals passing between the ships were eminently satisfactory. The door of the testing room is almost always shut, and the electricians pursue their work undisturbed; but it is impossible to exclude that spirit of inquiry which will satiate its thirst for scientific information through a keyhole.

Further the weather was all that could be wished for. Indeed, had the poet who was so anxious for 'life on the ocean wave and a home on the rolling deep' been abroad, he would have been absolutely happy, and perhaps even more desirous for a fixed habitation.

... Suddenly without an instant's warning, or the occurrence of any single incident that could account for it, the cable parted when subject, to a strain of less than a ton. (It was later discovered that the cable had been damaged during the storm) The gun that again told the *Valorous* of this fatal mishap brought all on board the *Agamemnon* rushing to the deck, for none could believe the rumour that had spread like wildfire about the ship. But there stood the machinery, silent and motionless, while the fractured end of the wire hung over the stern wheel, swinging loosely to and fro.

... It was evident then that the *Niagara* had rigidly, but most unfortunately, adhered to the mere letter of the agreement regarding the 100 miles and after the last fracture had at once turned back to Queenstown. On Tuesday the 6th, therefore, as the dense fogs and winds set in again it was agreed between the *Valorous* and *Agamemnon* to return once more to the rendezvous. But as usual the fog was so thick that the whole American navy might have been cruising there unobserved; so the search was given up, and at eight o'clock that night the ship's head was turned for Cork, and, under all sail, the *Agamemnon* at last stood homewards. The voyage home was made with ease and swiftness considering the lightness of the wind, the trim of the ship, and that she only steamed three days, and at midday on Tuesday, July 12th, the *Agamemnon* cast anchor in Queenstown Harbour, Ireland, having met with more dangerous weather, and encountered more mishaps than often falls to the lot of any ship in a cruise of thirty-three days.

Cyrus Field reported back to the Atlantic Telegraph Company in London, where several members, including chairman Sir William Brown, wanted to sell off the left-over cable and give over any profits to the shareholders. Some, however, wanted to make another attempt. Among them were Professor Thomson, Curtis Lampson (who was voted in as deputy chairman), and Stuart Wortley, who became chair after Sir William Brown resigned. The second attempt was made, without delay, just over a month later, from Queenstown on Saturday 17 July 1858.

Charles Bright's memoirs take up this second attempt:

As the ships left the harbour there was apparently no notice taken of their departure by those on shore or in the vessels anchored around them. Everyone seemed impressed with the conviction that we were engaged in a hopeless enterprise; and the squadron seemed rather to have slunk away on some discreditable mission than to have sailed for the accomplishment of a grand national scheme...

In consequence of continued delays and changes from steam to sail, and from sail to steam again, much fuel was expended, and not more than eighty miles of distance made good each day. On Sunday the 25th, however, the weather changed, and for several days in succession there was an uninterrupted calm. The moon was just at the full, and for several nights it shone with a brilliancy which turned the smooth sea into one silvery sheet, which brought out the dark hull and white sails of the ship in strong contrast to the sea and sky as the vessel lay all but motionless on the water, the very impersonation of solitude and repose. Indeed, until the rendezvous was gained, we had such a succession of beautiful sunrises, gorgeous sunsets, and tranquil moonlight nights as would have excited the most enthusiastic admiration of anyone, but persons situated as we were. But by us such scenes were regarded only as the annoying indications of the calm which delayed our progress and wasted our coals. To say that it was calm is not doing full justice to it; there was not a breath in the air, and the water was as smooth as a millpond. Even the wake of the ship scarce ruffled the surface; and the gulls which had visited us almost daily, and to which our benevolent liberality had dispensed innumerable pieces of pork, threw an almost unbroken shadow upon it as they stooped in their flight to pick up the largest and most tempting. It was generally remarked that cable laying under such circumstances would be mere child's play.

In spite of the unusual calmness of the weather in general, there were days on which our former unpleasant experiences of the Atlantic were brought forcibly to our recollection, when it blew hard and the sea ran sufficiently high to reproduce on a minor scale some of the discomforts of which the previous cruise had been so fruitful. Those days, however, were the exception and not the rule, and served to show how much more pleasant was the inconvenient calm than the weather which had previously prevailed.

After the ordinary laconic conversation which characterise code flag signals, we were as usual greeted by a perfect storm of questions as to what had kept us so much behind our time, and learned that all had come to the conclusion that the ship must have got on shore on leaving Queenstown [Ireland] harbour. The *Niagara*, it appeared, had arrived at the rendezvous on Friday night, the 23rd, the *Valorous* on Sunday, the 25th, and the *Gorgon* on the afternoon of Tuesday, the 27th.

The day was beautifully calm, so no time was to be lost before making the splice in lat. 52° 9′ N., long. 32° 27′ W., and soundings of 1500 fathoms. Boats were soon lowered from the attendant ships; the two vessels made fast by a hawser, and the *Niagara's* end of the cable conveyed on board the *Agamemnon*. About half past twelve o'clock the splice was effectually made, but with a very different frame from the carefully rounded semi-circular boards which had been used to enclose the junctions on previous occasions. It consisted merely of two straight boards hauled over the joint and splice, with the iron rod and leaden plummet attached to the centre. In hoisting it out from the side of the ship, however, the leaden sinker broke short off and fell overboard. There being no more convenient weight at hand a 32 lb. shot was fastened to the splice instead and the whole apparatus was quickly dropped into the sea without any formality and, indeed, almost without a spectator, for those on board the ship had witnessed so many beginnings to the telegraphic line that it was evident they despaired of there ever being an end to it.

The stipulated 210 fathoms of cable having been paid out to allow the splice to sink well below the surface, the signal to start was hoisted, the hawser cut loose, and the *Niagara* and *Agamemnon* start for the last time at about 1 pm for their opposite destinations.

The announcement comes from the electrician's testing room that the continuity is perfect, and with this assurance the engineers go on more boldly with the work. In point of fact the engineers may be said to be very much under the control of the electricians during paying out; for if the latter report anything wrong with the cable, the engineers are brought to a stand until they are allowed to go on with their operations by the announcement of the electricians that the insulation is perfect and the continuity all right. The testing room is where the subtle current which flows along the conductor is generated, and where the mysterious apparatus by which electricity is weighed and measured— as a marketable commodity—is fitted up. The system of testing and of transmitting and receiving signals through the cable from ship to ship during the process of paying out must now be briefly referred to. It consists of an exchange of currents sent alternately every ten minutes by each ship. These not only serve to give an accurate test of the continuity and insulation of the conducting wire from end to end, but also to give certain signals which it is desirable to send for information purposes. For instance, every ten miles of cable paid out is signalised from ship to ship, as also the approach to land or momentary stoppage for splicing, shifting to a fresh coil, etc. The current in its passage is made to pass through an electromagnetometer, an instrument invented by Mr Whitehouse. It is also conveyed in its passage at each end of the cable through the reflecting galvanometer and speaking instrument just invented by Prof. Thomson; and it is this latter which is so invaluable, not only for the interchange of signals, but also for testing purposes. The deflections read on the galvanometer, as also the degree of charge and discharge indicated by the magnetometer, are carefully recorded. Thus, if a defect of continuity or insulation occurs it is brought to light by comparison with those received before.

Shortly after four o'clock a very large whale was seen approaching the starboard bow at a great speed, rolling and tossing the sea into foam all round; and for the first time we felt a possibility for the supposition that our second mysterious breakage of the cable might have been caused after all by one of these animals getting foul of it under water. It appeared as if it were making direct for the cable; and great was the relief of all when the ponderous living mass was seen slowly to pass astern, just grazing the cable where it entered the water—but fortunately without doing any mischief. All seemed to go well to about eight o'clock; the cable paid out from the hold with an evenness and regularity which showed how carefully and perfectly it had been coiled away. The paying out machine also worked so smoothly that it left nothing to be desired. The brakes are properly called self-releasing; and although they can, by means of additional weights, be made to increase the pressure or strain upon the cable, yet, until these weights are still further increased (at the engineer's instructions), it is impossible to augment the strain in any other way. To guard against accidents which might arise in consequence of the cable having suffered injury during the storm, the indicated strain upon the dynamometer was never allowed to go beyond 1700 lb. or less than one quarter what the cable is estimated to bear. Thus far everything looked promising.

But in such a hazardous work no one knows what a few minutes may bring forth, for soon after eight o'clock an injured portion of the cable was discovered about a mile or two from the portion paying out. Not a moment was lost by Mr Canning, the engineer on duty, in setting men to work to cobble up the injury as well as time would permit, for the cable was going out at such a rate that the damaged portion would be paid overboard in less than twenty minutes, and former experience had shown us that to check either the speed of the ship or the cable would in all probability, be attended by the most fatal results, Just before the lapping was finished, Professor Thomson reported that the electrical continuity of the wire had ceased, but that the insulation was still perfect. Attention was naturally directed to the injured piece as the probable source of the stoppage, and not a moment was lost in cutting the cable at that point with the intention of making a perfect splice.

To the consternation of all, the electrical tests applied showed the fault to be overboard, and in all probability some fifty miles from the ship.

Not a second was to be lost, for it was evident that the cut portion must be paid overboard in a few minutes; and in the meantime, the tedious and difficult operation of making a splice had to be performed. The ship was immediately stopped, and no more cable paid out than was absolutely necessary to prevent it breaking. As the stern of the ship was lifted by the waves a scene of the most intense excitement followed. It seemed impossible, even by using the greatest possible speed, and paying out the least possible amount of cable, that the junction could be finished before the part was taken out of the hands of the workmen. The main hold presented an extraordinary scene. Nearly all the officers of the ship and of those connected with the expedition stood in groups about the coil, watching with intense anxiety the cable as it slowly unwound itself nearer and nearer the joint, while the workmen worked at the splice as only men could work who felt that the life and death of the expedition depended upon their rapidity. But all their speed was to no purpose, as the cable was unwinding within a hundred fathoms; and, as a last and desperate resource, the cable was stopped altogether, and for a few minutes the ship hung on by the end. Fortunately, however, it was only for a few minutes, as the strain was continually rising above two tons and it would not hold on much longer. When the splice was finished, the signal was made to loose the stoppers, and it passed overboard in safety.

When the excitement, consequent upon having so narrowly saved the cable, had passed away, we awoke to the consciousness that the case was yet as hopeless as ever, for the electrical continuity was still entirely wanting.

Preparations were consequently made to pay out as little rope as possible, and to hold on for six hours in the hope that the fault, whatever it was, might mend itself, before cutting the cable and returning to the rendezvous to make another splice. The magnetic needles on the receiving instruments were watched closely for the returning signals, when, in a few minutes, the last hope was extinguished by their suddenly indicating dead earth, which tended to show that the cable had broken from the *Niagara,* or that the insulation had been completely destroyed.

Nothing, however, could be done. The only course was to wait until the current should return or take its final departure. And it *did* return—with greater strength than ever—for in three minutes everyone was agreeably surprised by the intelligence that the stoppage had disappeared and that the signals had again appeared at their regular intervals from the *Niagara*. It is needless to say what a load of anxiety this news removed from the minds of everyone, but the general confidence in the ultimate success of the operations was much shaken by the occurrence, for all felt that every minute a similar accident might occur.

During the morning of Friday the 30th, everything went well. The ship had been kept at the speed of about five knots, the cable going out at six, the average angle with the horizon at which it left the ship being about 15°, while the indicated strain upon the dynamometer seldom showed more than 1600 lb. to 1700 lb.

During the evening, top masts were lowered, and spars, yards, sails and indeed everything aloft that could offer resistance to the wind, was sent down on deck. Still the ship made but little way, chiefly in consequence of the heavy sea, though the enormous quantity of fuel consumed showed us that if the wind lasted, we should be reduced to burning the masts, spars, and even the decks, to bring the ship into Valentia [Ireland]. It seemed to be our particular ill fortune to meet with head winds whichever way the ship's head was turned. On our journey out we had been delayed and obliged to consume an undue proportion of coal for want of an easterly wind, and now all our fuel was wanted *because* of one. However, during the next day the wind gradually went round to the southwest, which, though it raised a very heavy sea, allowed us to husband our small remaining store of fuel.

At noon on Saturday, the 31st of July, observations showed us to be in lat. 52° 23′ N., and long. 26° 44′ W. having made good 120 miles of distance since noon of the previous day, with a loss of about 27 per cent. of cable. The *Niagara*, as far as could be judged from the amount of cable she paid out —which by a previous arrangement was signalled at every ten miles—kept pace with us, within one or two miles, the whole distance across.

Now, indeed, were the energy and activity of all engaged in the operation tasked to the utmost. Mr Hoar and Mr Moore— the two engineers who had the charge of the relieving wheels of the dynamometer—had to keep watch alternately every four hours, and while on duty durst not let their attention be removed from their occupation for one moment; for on their releasing the brakes every time the stern of the ship fell into the trough of the sea entirely depended the safety of the cable, and the result shows how ably they discharged their duty.

Throughout the night there were few who had the least expectation of the cable holding on till morning, and many lay awake listening for the sound that all most dread to hear, viz., the gun which should announce the failure of all our hopes. But still the cable—which in comparison with the ship from which it was paid out, and the gigantic waves among which it was delivered, was but a mere thread—continued to hold on, only leaving a silvery phosphorous line upon the stupendous seas as they rolled on towards the ship.

At noon on Monday, August the 2nd, observations showed us to be in lat. 52° 35′ N., long. 19° 48′ W. Thus, we had made good 127 miles since noon of the previous day and had completed more than halfway to our ultimate destination.

During the afternoon, an American three masted schooner, which afterwards proved to be the *Chieftain* was seen standing from the eastward towards us. No notice was taken of her at first, but when she was within about half a mile of the *Agamemnon*, she altered her course and bore right down across our bows. A collision which might prove fatal to the cable now seemed inevitable; or could only be avoided by the equally hazardous expedient of altering the *Agamemnon's* course. The *Valorous* steamed ahead and fired a gun for her to heave to, which, as she did not appear to take much notice of, was quickly followed by another from the bows of the *Agamemnon,* and a second and third from the *Valorous.* But still the vessel held on her course; and, as the only resource left to avoid a collision, the course of the *Agamemnon* was altered just in time to pass within a few yards of her. It was evident that our proceedings were a source of the greatest possible astonishment to them, for all her crew crowded upon her deck and rigging. At length they evidently discovered who we were and what we were doing, for the crew manned the riggings, and, dipping their ensign several times, they gave us three hearty cheers. Though the *Agamemnon* was obliged to acknowledge these congratulations in due form, the feeling of annoyance with which we regarded the vessel—which (either by the stupidity or carelessness of those on board) was so near adding a fatal and unexpected mishap to the long chapter of accidents which had already been encountered— may easily be imagined.

To those below—who of course did not see the ship approaching—the sound of the first gun came like a thunderbolt, for all took it as a signal of the breaking of the cable. The dinner tables were deserted in a moment, and a general rush made up the hatches to the deck; but before reaching it their fears were quickly banished by the report of the succeeding gun, which all knew well could only be caused by a ship in our way, or a man overboard.

Throughout the greater part of Monday morning, the electrical signals from the *Niagara* had been getting gradually weaker, until they ceased altogether for nearly three quarters of an hour. Then Professor Thomson sent a message to the effect that the signals were too weak to be read; and, in a little while, the deflections returned even stronger than they had ever been before. Towards the evening, however, they again declined in force for a few minutes. [Note: *It subsequently transpired that the trouble had been due to a fault in the Niagara's wardroom coil. As soon as the electricians discovered this, and had it cut out, all went smoothly again.]*

With the exception of these little stoppages, the electrical condition of the submerged wire seemed to be much improved. It was evident that the low temperature of the water at the immense depth improved considerably the insulating properties of the gutta percha, while the enormous pressure to which it must have been subjected probably tended to consolidate its texture, and to fill up any air bubbles or slight faults in manufacture which may have existed.

The weather during Monday night moderated a little; but still there was a very heavy sea on, which endangered the wire every second minute.

About three o'clock on Tuesday morning all on board were startled from their beds by the loud booming of a gun. Everyone—without waiting for the performance of the most particular toilet—rushed on deck to ascertain the cause of the disturbance. Contrary to all expectation, the cable was safe; but just in the grey light could be seen the *Valorous* rounded to in the most warlike attitude— firing gun after gun in quick succession towards a large American barque, which, quite unconscious of our proceedings, was standing right across our stern. Such loud and repeated remonstrances from a large steam frigate were not to be despised; and evidently without knowing the why or the wherefore she quickly threw her sails aback and remained hove to. Whether those on board her considered that we were engaged in some filibustering expedition, or regarded our proceedings as another outrage upon the American flag it is impossible to say; but certain it is that apparently in great trepidation—she remained hove to until we had lost sight of her in the distance.

Tuesday was a much finer day than any we had experienced for nearly a week, but still there was a considerable sea running, and our dangers were far from past; yet the hopes of our ultimate success ran high. We had accomplished nearly the whole of the deep portions of the route in safety, and that, too, under the most unfavourable circumstances possible; therefore, there was every reason to believe that unless some unforeseen accident should occur we should accomplish the remainder. Observations at noon placed us in lat. 52° 26′ N., long. 16° 7′ 40″ W., having run 134 miles since the previous day.

About five o'clock in the evening the steep submarine mountain which divides the steep telegraphic plateau from the Irish coast was reached, and the sudden shallowing of water had a very marked effect upon the cable, causing the strain and the speed to lessen every minute. A great deal of slack was paid out, to allow, for any greater inequalities which might exist, though undiscovered by the sounding line. [Note: *The amount of slack paid out had already been almost ruinous. Luckily its continuance was not necessary, or it would have been impossible to reach Ireland with the cable on board.*]

About ten o'clock the shoal water of 250 fathoms was reached. The only remaining anxiety now was the changing from the lower main coil to that upon the upper deck; and this most dangerous operation was successfully performed between three and four o'clock on Wednesday morning.

Wednesday was a beautiful, calm day. At noon we were in lat. 52° 11′; long. 12° 40′ 2″ W., eighty-nine miles distant from the telegraph station at Valentia. The water was shallow, so that there was no difficulty in paying out the wire almost without any loss by slack; and all looked upon the undertaking as virtually accomplished.

By daylight on the morning of Thursday, the 5th, the bold, rocky mountains which entirely surround the wild and picturesque neighbourhood of Valentia rose right before us at a few miles distance. Never, probably, was the sight of land more welcome, as it brought to a successful termination one of the greatest, but at the same time most difficult, schemes which was ever undertaken. Had it been the dullest and most melancholy swamp on the face of the earth that lay before us, we should have found it a pleasant prospect; but as the sun rose behind the estuary of Dingle Bay, tingeing with a deep, soft purple, the lofty summits of the steep mountains which surround its shores illuminating the masses of morning vapour which hung upon them, it was a scene which might vie in beauty with anything that could be produced by the most florid imagination of an artist.

No one on shore was apparently conscious of our approach, so the *Valorous* went ahead to the mouth of the harbour and fired a gun. Both ships made straight for Doulas Bay, the *Agamemnon* steaming into the harbour with a feeling that she had done something, and about 6 am came to anchor at the side of Beginish Island, opposite to Valentia.

As soon as the inhabitants became aware of our approach, there was a general desertion of the place, and hundreds of boats crowded round us—their passengers in the greatest state of excitement to hear all about our voyage. The Knight of Kerry was absent in Dingle, but a messenger was immediately despatched for him, and he soon arrived in Her Majesty's gunboat *Shamrock*.

Soon after our arrival a signal was received from the *Niagara* that they were preparing to land, having paid out 1030 nautical miles of cable, while the *Agamemnon* had accomplished her portion of the distance with an expenditure of 1020 miles, making the total length of the wire submerged 2050 geographical miles.

Immediately after the ships cast anchor, the paddle box boats of the *Valorous* were got ready, and two miles of cable coiled away in them, for the purpose of landing the end. But it was late in the afternoon before the procession of boats left the ship, under a salute of three rounds of small arms from the detachment of marines on board the *Agamemnon*, under the command of Lieutenant Morris.

The progress of the end to the shore was very slow, in consequence of the stiff wind which blew at the time; but at about 3 pm the end was safely brought on shore at Knight's Town, Valentia, by Mr Bright, to whose exertions the success of the undertaking is attributable. Mr Bright was accompanied by Mr Canning and the Knight of Kerry. The end was immediately laid in the trench which had been dug to receive it; while a royal salute, making the neighbouring rocks and mountains reverberate, announced that the communication between the Old and New World had been completed.

The cable was taken into the electrical room by Mr Whitehouse, and attached to a galvanometer, and the first message was received through the entire length now lying on the bed of the sea.

Too much praise cannot be bestowed upon both the officers and men of the *Agamemnon* for the hearty way in which they have assisted in the arduous and difficult service they have been engaged in; and the admirable manner in which the ship was navigated by Mr Moriarty materially reduced the difficulty of the Company's operations.

It will, in all probability, be nearly a fortnight before the instruments are connected at the two termini for the transmission of regular messages.

It is unnecessary here to expatiate upon the magnitude of the undertaking which has just been completed, or upon the great political and social results which are likely to accrue from it; but there can be but one feeling of universal admiration for the courage and perseverance which have been displayed by Mr Bright, and those who acted under his orders, in encountering the manifold difficulties which arose on their path at every step.

Niagara journeyed to Newfoundland without a problem, apart from a deficiency in the wardroom coil, which was located and cut out. She arrived in Trinity Bay, where HMS *Porcupine* was waiting to give aid if needed. *Niagara* put down her anchor at 1 a.m. on 5 August and that afternoon the cable uncoiled in Bull Arm Bay, which is at the tip of Trinity Bay.

Charles Bright communicated with the board in London at once. They then passed on his message to the press:

Charles Bright, to the Directors of the Atlantic
Telegraph Company.
VALENTIA, Aug. 5th.
The *Agamemnon* has arrived at Valentia, and we are about to land the end of the cable.

The *Niagara is* in Trinity Bay, Newfoundland. There are good signals between the ships.

We reached the rendezvous on the night of the 28th, and the splice with the *Niagara* cable was made on board the *Agamemnon* the following morning.

By noon on the 30th 265 nautical miles were laid between the ships; on the 31st, 640; on the 1st August, 884; on the 2nd, 1256; on the 4th, 1854; on anchoring at six in the morning in Doulas Bay, 2022.

The speed of the *Niagara* during the whole time has been nearly the same as ours, the length of cable paid out from the two ships being generally within ten miles of each other.

With the exception of yesterday, the weather has been very unfavourable.

The first successful correspondence arrived from Newfoundland on 13 August 1858, with a reply from Ireland being sent on 16 August:

Europe and America are united by telegraphy. Glory to God in the highest, on earth peace, goodwill towards men.

The queen was quick to pass on her thoughts:

> From Her Majesty the Queen of Great Britain to His Excellency the President of the United States.
>
> The Queen desires to congratulate the President upon the successful completion of this great international work, in which the Queen has taken the greatest interest.
>
> The Queen is convinced that the President will join with her in fervently hoping that the electric cable, which now already connects Great Britain with the United States, will prove an additional link between the two nations, whose friendship is founded upon their common interest and reciprocal esteem.
>
> The Queen has much pleasure in thus directly communicating with the President, and in renewing to him her best wishes for the prosperity of the United States.

President James Buchanan replied:

> WASHINGTON CITY.
> The President of the United States to Her Majesty Victoria, Queen of Great Britain.
>
> The President cordially reciprocates the congratulations of Her Majesty the Queen on the success of the great international enterprise accomplished by the skill, science, and indomitable energy of the two countries.
>
> It is a triumph more glorious ... far more useful to mankind than was ever won by a conqueror on the field of battle.
>
> May the Atlantic Telegraph, under the blessing of heaven, prove to be a bond of perpetual peace and friendship between the kindred nations, and an instrument destined by Divine Providence to diffuse religion, civilisation, liberty, and law throughout the world.
>
> In this view will not all the nations of Christendom spontaneously unite in the declaration that it shall be for ever

neutral, and that its communication shall be held sacred in passing to the place of their destination, even in the midst of hostilities?

JAMES BUCHANAN

The cable was used to send a variety of messages. For instance, the British commanding officers at Halifax and Montreal were sent orders refuting previous commands that had been sent by mail. These new orders were for the 39th and 62nd regiments to make their way back to England, saving the British government £50,000 to £60,000. The cable would be used to send 732 communications before it broke on 20 October 1858. Charles Bright was honoured with a knighthood, and the Gutta Percha Company gained immense expertise.

Communication between the two countries had been achieved, but only at a rate of three words per minute and only until the cable broke. Professor Thomson blamed Whitehouse for its failure. Thomson pursued the idea of the mirror galvanometer, which he thought was a machine more capable of discovering electrical current rather than increasing voltage (up to 2,000 volts), as Whitehouse did. In Thomson's view, this damaged the insulation. Of course, he was right, and it was very beneficial that he was involved in the experiments. Thomson knew that the usefulness of copper as a conductor was reliant on its freedom from contamination and wanted to include a quality control in subsequent cable projects.

Cyrus Field was eager to try yet again, and started fundraising. This time, British industry was more cautious, while Americans, as Field's brother observes, 'passed a series of resolutions, in which they applauded the projected telegraph across the ocean as one of the grandest enterprises ever undertaken by man, which they proudly commended to the confidence and support of the American public ... But not a man subscribed a dollar.'

The winds of time blew events in a different direction. In 1858, the UK was ending its conflict with Russia but there were problems with China and mutinies in India. France was at war as well, invading Austria, and although the USA was at peace it would not last. The American Civil War soon took over, while Britain – and indeed George – began to turn its attention to the Suez Canal. Darwin wrote *On the Origin of Species*, and Cardinal Newman wrote his bestseller *Apologia Pro Vita Sua*, or *A Defence of One's Own Life*, in which he attacked the Protestant Church. London saw the first trains run on the Underground, as well as hosting the Great London Exposition in 1862. The world had averted its gaze, and it was not until the end of the American Civil War that another cable was attempted.

In that time the Gutta Percha Company manufactured over thirty cables for the North Sea, the Mediterranean, the Irish Sea, the Black Sea, the Indian Ocean and the fjords in Scandinavia. Other projects included cables for the Red Sea, the Aegean, the English Channel, the Pacific, the Adriatic, the Tasman Sea, the South Atlantic, the Arabian Sea, the Persian Gulf and the Bay of Bengal. The company was busy, with these projects involving 14,000 nautical miles of cable.

The British government, understandably, did not want to be left behind and so formed a joint committee to get to the bottom of why the Transatlantic Cable had not proved a lasting success. Over a series of twenty-two meetings it was decided that the gutta percha was not efficient enough to use as an insulator. Willoughby Smith had very strong views on this, and he managed to persuade the committee that the wire core should be soaked in a conducting fluid rather than the gutta percha, which was an insulator. He went on to conclude that the core should be kept in a tank full of water so that its electrical capacity could be observed without interruption. However, gutta percha continued to be utilised, with imports to the UK reaching 1,000 tons by 1861.

Glass, Elliot and Company soon took on Sir Samuel Canning and Henry Clifford. Richard Glass was certain that the only way to conquer the massive task of laying a cable under the Atlantic was to have the project administered by one company. With this in mind, the Gutta Percha Company and Glass, Elliot and Company joined forces on 7 April 1864 to become the Telegraph Construction and Maintenance Company, otherwise known as Telcon. They were keen to make another attempt at the Transatlantic cable and approached Cyrus Field and his Atlantic Telegraph Company. An agreement was reached, and they unified for the purpose of the venture.

George Elliot was a member of the board of Telcon until his resignation in 1869, when his eldest son, Ralph, took his place while he busied himself with his parliamentary career. Ralph was to work there for three years before resigning to make room for George to return to the board.

One hundred years after their inception, Telcon would proudly announce that they had made 82 percent of all the world's submarine cables, amounting to 385,000 nautical miles. In George's time, until his death in 1894, they made cables for prestigious clients. Among them were jobs in the early 1870s involving 3,268 nautical miles for the Eastern Telegraph Co., laid between Suez, Aden and Bombay; 3,180 nautical miles for the same company, linking Cornwall, Gibraltar, Malta and Alexandria; and 2,366 nautical miles for the Eastern Extension Telegraph Co., linking Madras, Penang and Singapore. As can be guessed from the locations involved, these three cables were utilised to unite the British Empire. Further links to the empire followed in 1876, with a 1,283-naut cable laid for the Eastern Extension Company, linking Australia and New Zealand, and two more for the same company the following year, totalling 2,742 nautical miles, linking Rangoon and Penang and the East Coast of Africa – running through Durban, Delagoe, Mozambique, Zanzibar and Aden. This was followed by a cable made for the African Direct Telegraph Co., which extended the empire cable links by joining the west coast of Africa from Bathurst, Gambia to Cape Town, South Africa with a 2,078-nautical-mile cable. Further links were made with a

2,679-nautical-mile cable that same year in South Africa, and from 1889 to 1901 over 6,885 nautical miles of cable were laid between Cornwall and Cape Town, stopping on islands on the west coast of Africa, enabling communication during the Boer War. Today, Telcon continues to be a world leader in soft magnetic cores and components.

Wire rope manufacture was a huge market in the mid-1850s, completely outstripping all the other wire trades. Enormous quantities were required by the Royal Navy, shipping merchants, the coal trade and other underground excavations being performed overseas. The railway network was another large consumer, and businesses which were in the thrust of the Industrial Revolution all needed wire rope in order to perform. There was a point when the manufacture of submarine telegraph cables was more dominant than all the other markets put together.

All copper cables, from 1851 to the 1980s, when they were replaced by fibreoptic, were made in the same way, with three main components. First was the copper core, which allowed the signal to pass; then the gutta percha insulation, which stopped water from ruining the signal; and finally the iron or steel armouring wires, which lent strength to the wire and protected it near the shore. In the 1850s, the completed cable was created by several different businesses. The wire-drawing companies came first, making thousands of miles of copper wire. Work was then taken over by the Gutta Percha Company, who fabricated the insulation before handing over to a third company, such as Glass, Elliot and Company, who would finish the wire by spinning the outer armouring wires – in turn supplied by another business like Webster and Horsfall – around the core to produce the finished item.

There were three main wire rope producers in the UK: Glass, Elliot and Company of Cardiff, Newall and Company of Gateshead, and Webster and Company of Sunderland. The biggest of these was Glass, Elliot and Company, predominantly because of their involvement with the mines and their rapid reaction and response to producing submarine cables.

The Transatlantic Cable was something out of a storybook, such was the excitement and allure of joining the two continents. The press bought emotions to a fever pitch. What had been made obvious by the first attempt at laying the cable in 1858 using the battleship *Agamemnon* was that the iron cladding around the wire was simply not up to the job of protecting it, even though the sheer weight of the cable put the ship in extreme peril.

By 1862 the Gutta Percha Company was well established, having produced more than forty-four underwater wires in use around the globe. They were effective, and although repairs were necessary at the ends, where they were damaged by falling anchors, none needed mending at any other point. Glass, Elliot and Company were learning from their experiences, and in keeping with their imaginative and dynamic nature they put forward a new endeavour to Field in March 1863. They wanted to manufacture a new transatlantic wire at the cost of £700,000, 20 per cent of which was to be paid by the Atlantic Telegraph Company through the sale of shares.

Glass, Elliot and Company were the most resourceful and inventive cable makers of the day. Cyrus Field was so excited about this prospect that he went direct to the wire manufacturers, Webster and Horsfall, without talking it through with Glass, Elliot and Company. Webster and Horsfall did not need the business as they were receiving requests from all over the Continent. However, things were going to sour in their relationship with Glass, Elliot and Company, who they disliked for various reasons, one being they 'could not even keep their own men together'.

Richard Glass and George Elliot parted company before the 1865 cable was completed. They did not give notice to their business associates, and their commercial administrator, Thomas Nixon, who was dealing with Webster and Horsfall, resigned. He asked advice from Webster and Horsfall about joining Glass, Elliot and Company's rivals, London-based Wilkins and Weatherly, and took a job with them.

Field was after 1,000 tons of wire from Webster and Horsfall at a time when, in 1863, they were only able to produce 400 tons a year. However, the company priced the order at £53 10s 8d per ton, mentioning to Field their disquiet at having to deal with Glass, Elliot and Company. James Horsfall commented that it would be an impossible task, not just for his company but for any in Great Britain. However, he did add, 'If I could have the order ensured to me, I could undertake to execute it in about twelve months but to do that I must make extensive additions to my works.'

A letter to Glass, Elliot and Company from Thos. Firth and Sons, steel manufacturers in Sheffield, offers advice to the company:

We think it will be absolutely necessary for you to give more time, if you are to receive the whole supply from one maker which we most decidedly advise you to do. It might be obtained from several Houses by the time you specify but if you want to secure uniformity and success you must have it from one Maker and one uniform quality and temper. This we consider most important and essential to the success of this Cable and there is no other House in this Country that could supply it in the same time as Messrs Webster and Horsfall.

Cyrus Field received the same advice, leading to his tender to Webster and Horsfall. James Horsfall had to cancel all existing business arrangements to make way for this huge commission, and was worried:

I am convinced in my own mind that success is very *doubtful*; and the disgrace – must fall to the lot of someone or other *which would no doubt be published to the World,* knowing that Webster and Horsfall have a reputation at stake.

He communicated to Richard Glass in the same manner:

> I shall call at your office tomorrow at 11 o'clk and be
> prepared to sign the enclosed offer, but before I do so, it will
> be necessary for your firm to sign the enclosed authority; as
> our agreement binds us to apply ¾ of the capabilities of our
> works in executing the orders of Glass, Elliot and Co. If we
> accept this order, we must be at liberty to use the remaining
> ¼th as we choose, otherwise we should entirely ruin our
> connection.

James Horsfall signed the contract for the 1865 Transatlantic
Cable on 9 May 1864. His firm was to provide 1,600 tons of
0.095 crucible cast steel wire, which was without a patent, and had
a tensile strength in the region of 70 tons for every square inch.
James legally promised to provide the wire within a year, even
though his mill was at this time incapable of producing even half
the quantity. He was as good as his word, however, and finished
the work early, in March 1865.

James made sure he met either with Cyrus Field, Glass or
Canning in London every week. He in turn put pressure on the
Chillingworth Iron Company, which was providing him with raw
materials. Tensions were high, straining relationships. He wrote to
Richard Glass on 14 June 1864, saying,

> In a note from Mr Elliot he tells me that the business belongs
> solely to him. Will you have the kindness to tell me in which
> business you are now his partner – in fact which is your
> business, and which is his. As my engagements are very
> large with you it is right I should know with whom I am in
> future to deal. Be pleased to say *when* the dissolution was
> advertised.

This split in Glass, Elliot and Company was unimaginable when
there was so massive a contract in place, and the fact that James
did not hear about it first-hand must have really angered him.

Not much is known about the spat that led to the breaking of partnership between Glass and Elliot. James Horsfall penned a letter to George regarding a visit to his works:

I am sorry to say I cannot meet your wishes and allow an inspection for I never allow anyone to enter our works. I am sorry to refuse this, but it is a rule which was deemed absolutely necessary by Mr Webster and I see the advisability of following it strictly.

This must have incensed George, especially in light of the fact that the Earl and Countess of Caithness had a guided tour of Hay Mills shortly afterwards. James had built the New Mill specially to cope with the contract for the 1865 Transatlantic Cable. But evidently it was felt at Horsfall and Webster that 'Sir George Elliot ... seems to have been one of a brand type born of the Victorian era.' His manner found ill favour among the staff at Horsfall and Webster:

In our time men without principle have had too free a rein. Even the permissive society has stretched its evil hand in to business. The take-over or merger is so often only another conspiracy, and what shoddy proceedings most of them are. A few investors may find such adventures amusing but perhaps they have not met any of the staff who have suffered in the process. The desperate search for power, or for wealth, unrefined by worthier considerations, so rarely brings happiness. I am thankful that no one among W.H. owners was ever afflicted by such motives. Whatever we may have done, our members never failed in that way and they were always guided by impulses which were honourable, and which were usually unselfish.

James was firm. He wrote to George on 9 July 1864:

As your partnership with Mr Glass is dissolved and you are now the sole owner ... our agreement of 24 March 1857 is terminated. If you wish to enter in to a new one with altered terms we think there will be no difficulty...

James had largely disposed of his other contracts just to concentrate on making the cable. Life was extremely stressful, and when one of the engines at the works broke down, destroying several wheels and putting the machine out of action for ten to fifteen days at a repair cost of over £1,000, Elliot's chief engineer, Canning, wrote to James saying, '*YOU* will prevent the Atlantic cable being laid in 1865 if you do not look out.'

Webster and Horsfall concluded their work on 1 May 1865, having completed an additional 67 tons, the order for which was placed on 6 April 1865. Cyrus Field was invited on a tour of the factory at Killamarsh and also visited their iron supplier. He responded by taking James to see the SS *Great Eastern*, the ship that would lay the cable. He wrote,

> The *Great Eastern* was the only steam ship in the world with the capacity to hold the 1865 Cable, which was substantially larger and heavier than its predecessor; this was mainly due to the seven wire stranded copper conductor which was treble the capacity of the solid conductor of 1858. Although the distance from Valentia on the west coast of Ireland to Trinity Bay, Newfoundland, was only 1,600 miles, in fact 2,110 miles had been expended in the laying of the 1858 Cable and 2,300 miles had been allowed for this time.

The SS *Great Eastern* launched in 1858. It was an iron sailing steamship designed by Isambard Kingdom Brunel, capable of carrying 4,000 passengers from England to Australia without stopping for fuel. The longest ship of its day at 692 feet, it retained that title until 1901. Brunel loved the ship, calling it the 'Great Babe'. Sadly, he died shortly after the maiden voyage. It had since been converted to a cable-laying ship, and now came the time for the 1865 cable to be laid.

The *Great Eastern* left Valentia on 23 July 1865, captained by James Anderson, carrying 7,000 tons of cable, 2,000 tons of material used to restrain it and a further 8,500 tons of coal, making a grand total of 21,000 tons. The mission quickly ended up disaster. On 2 August the cable broke, and although three attempts were made to hook the cable and bring it back to the surface, the 2-mile depth of the ocean proved insurmountable.

A repeat order was put to Webster and Horsfall, for 1,000 tons of steel wire. The idea was to lay a new cable, as well as retrieving the lost cable, which would be spliced and taken to Newfoundland. £500,000 was needed, which the British public provided, on shares that paid a dividend of 12 per cent.

The *Great Eastern* made her second attempt on 30 June 1866, taking fourteen days to accomplish the desired result. Her course ran 20 miles north of her attempt the previous year. Canning received a knighthood and Cyrus Field wrote enthusiastically to James Horsfall, 'Sir Samuel Canning will require the wire for the cable to India … I assure you I shall not fail to recommend your wire most warmly.'

What was achieved on 27 July 1866 was a permanent electrical communication between the two continents, with huge implications. Five more cables were laid up until 1894, linking the ports of Heart's Content in Newfoundland and Valentia in Ireland. They were still in operation in 1965.

But all was not well. Horsfall and Webster were unhappy with their business relationship with George, as Horsfall explains:

With Elliot, the situation was never again easy or cordial, not is it hard to understand why. George Elliot himself found difficulty in brooking an equal, and he surged through life with imperious disregard for the feelings of his fellow men. His partnership with Glass had not survived the stresses of 1864 and James had thrown in his lot with his erstwhile colleague. It seems likely that George Elliot always had a chip on his shoulder, certainly so from that time, and incidents such as James refusing him access to the works would not have helped ... This was the background to the unpleasantness and intrigue of the eighties which ended the connection for ever.

Later, in 1867, George presented a paper to the Institute of Mechanical Engineers describing the paying-out and picking-up machinery employed in the successful laying of the 1866 Atlantic telegraph cable:

The Atlantic Telegraph Cable expedition of 1866 was twofold in its purpose, the first object being to lay a new cable, and the second to recover and complete the one commenced and lost in the unsuccessful attempt of the previous year. The general arrangements on board the *Great Eastern* steamship, which was employed for the purpose on both occasions, were similar in both years; but the knowledge and experience which had been so dearly bought in 1865 rendered necessary many improvements and alterations in the detail of the apparatus employed.

The cable itself was coiled in three circular wrought-iron tanks, which were built on the main deck of the ship. The foremost tank A occupied the space which had previously been the forecargo space; and the after tank C was placed in what had been the aftercargo space. The middle tank B occupied what had been the second dining saloon; and the funnel from the pair of boilers in that position was removed for the purpose, those boilers being thrown out of work during the expedition. The whole of the fittings in these spaces had been removed, and each of the tanks was stayed to the sides of the ship by two flat frames of iron, built on angle-iron framing, thus securing the tanks in the most substantial manner. The deck had also been shored underneath by baulk timbers, which were carried through from deck to deck down to the bottom of the ship.

The fore tank was 51 ft. 6 ins. diameter, the middle tank 58 ft. 6 ins., and the after tank 58 ft. 0 ins.; they were all of a uniform depth of 20 ft. 6 ins., and similarly constructed in all respects. The bottoms were ½ inch thick, lap-jointed; and the sides were ⅝ inch thick in the lower half; and ½ inch in the upper half. The sides were butt-jointed, so as to present a perfectly smooth surface inside; and the bottoms were covered with a thin wood floor to receive the cable. As it was of vital importance that the cable should be kept always under water, to prevent depreciation of the gutta-percha coating, and also to afford the only means of effectually testing its electrical condition, these tanks were carefully made watertight. In paying out the cable, the water in the tanks was kept somewhat below the level of the top flake, and required to be lowered during the paying out; for this purpose each tank was supplied with discharge valves, and as the bottoms of the coils were above the water line of the ship, it was only necessary to open these valves in order to allow the tanks to discharge themselves completely.

The coiling of the cable into the tanks, out of the hulks by which it was brought from the Telegraph Construction and Maintenance Works at Greenwich to the *Great Eastern* at Sheerness, was affected in the following manner. The cable was brought up over the side of the ship from the hulk, upon wheels which guided it on to a large deep-grooved wheel driven by steam power; on the tread of this wheel ran a small jockey wheel or roller, pressing the cable down into the groove of the large wheel, so as to give sufficient friction for enabling the wheel to draw up the cable from the hulk. The coiling commenced from the outside of the tank, the end being previously triced up above the tank, leaving a clear end for splicing and testing. The first turn of the cable was carefully laid round the outside of the tank, and the next was laid back close up against the previous turn, and so on until a perfectly flat flake or layer was laid into the eye of the coil, which was left about 9 ft. 6 ins, diameter.

In paying out the cable it was passed up to the hatch over the centre of the coil and carried over a large wheel E about 4 feet diameter. The cable was then carried in a trough F about 2 feet wide, made of sheet iron, leading to the paying-out machinery; this trough was fitted with rollers at about 10 or 12 feet intervals to relieve the cable from friction in passing along, until it reached the paying-out machinery, which was placed in the stern of the ship, slightly to the port side.

The length of cable in the after tank was 840 knots (1 knot = 6084 feet = 1.15 statute mile), in the middle tank 865 knots, and in the fore tank 671 knots; and the entire length of 2376 knots was joined up into one continuous length of cable before the laying was commenced. The size of the cable was 1⅛ inch diameter, and its weight 31 cwts. (1.55 tons) per knot in air, and 14¾ cwts. (737.5kg) per knot when immersed in water; the breaking strain was 8.10 tons, equal to eleven times its weight in water per knot, so that the cable would just bear its own weight in 11 knots depth of water.

In the paying-out machinery the chief object to be attained was to supply some means of checking the cable in the most regular manner possible while passing out of the ship, and also of keeping it in a state of constant tension; and it was required that the amount of this tension should be at all times known, and that it should be regulated by the depth of the water in each particular part of the ocean, and also to some extent by the speed of the ship.

The most important feature is the arrangement by which it was rendered impossible that more than a certain strain should be kept upon the cable during the paying out. A less strain would only involve a slight loss of cable; but any increased strain might possibly damage or even destroy it. The cable on entering the paying-out machinery was passed over a series of six deep-grooved wheels, each about 3 feet diameter.

After the cable had passed through this part of the machinery, called for distinction the Jockey Gear, and had thereby been subjected to a slight amount of strain, it was led to the main Paying-Out Drum P.

The cable of 1865 had been laid with about 15 per cent. of slack, and this percentage of slack was the great source of hope for the successful recovery of the cable. It was calculated that if the cable could be raised to the surface without hooking it at more than a single point, there would be a bight [loop of rope] suspended in the water of 9¼ knots in length, when in 2 knots depth of water; and the horizontal distance would be 8 knots between the portions resting upon the ground, giving an excess of length of 15 per cent. in the suspended bight; and the results of the actual picking up proved this calculation to represent very closely the curve of the suspended cable. The size of the cable was 1⅛ inch diameter, and its weight 31 cwts. [1.55 tons] per knot in air, and 14 cwts. [0.7 tons] per knot when immersed in water; the total weight of a suspended length of 94 knots in water was therefore 6½ tons, but as the breaking strength of the cable was 7¾ tons, it would carry the weight of 11 knots of its own length in water before breaking. As however the possibility of its recovery in this manner in a single bight was generally considered to be out of the question, it was intended therefore to attempt raising it by degrees only. Three steamships were accordingly fitted with picking-up apparatus, the *Medway*, *Great Eastern*, and *Albany*, for the purpose of grappling for the cable simultaneously in three places: the *Medway* to grapple to the east and the *Albany* to the west of the *Great Eastern*.

Finding the lost 1865 cable had been a stroke of luck. It was 600 miles from shore, and 2.5 miles or 4 kilometres below the surface. Thirty attempts would be made before the team met with success. The two ships, *Medway* and *Albany*, got into position and *Albany* lowered her grappling hook, which was a five-pronged affair at the end of a rope. *Albany* had early success, finding the cable but it slipped from the buoy to which it had been attached overnight. Rough seas prevented the cable being secured in further attempts. It took two weeks for the cable to be fastened, and another twenty-six hours to load it on SS *Great Eastern*. The cable was spliced with a new one, held on the ship, and it was laid all the way to Heart's Content in Newfoundland, arriving on September 7, giving a second working Transatlantic Cable. Understandably, it was heralded as an even greater achievement than the successful laying of the 1866 cable.

George took what he had learnt on the Transatlantic Cable mission to Elliot and Co. This new company focused on the locked coil rope used in winding and haulage in mines, which had to be very strong. A report in 1889 suggests 'that 80 to 90 tons per sq. inch was the tensile strength decided upon in order to compete in breaking load with the existing round-strand ropes of that time'. Samples were cut and shown at the Science Museum, South Kensington and the Smithsonian Institute, Washington. They gained reputed excellence, being 'specially constructed so as to prevent twisting in ascending or descending'. The rope gained the approval of the mine inspectors, who noted that it 'did not twist or turn in the whole length of the shaft which is 750 yards deep', making the mines a much safer place.

By 1890, Elliot and Co. was in the business of making mile-long ropes for the mines. The rights to these ropes were subcontracted and sold to companies all over Europe and the United States.

One expert, T. H. Davies, in his biography of inventor Telford Clarence Batchelor, states:

> Broadly speaking the records appear to indicate that without the powerful support of Sir George Elliot, it is highly doubtful whether the invention would have been persisted in. Such are the trials and difficulties that, contrary to popular belief, occur in bringing master inventions to fruition.

They certainly made a huge difference in the safety of the mines.

5

THE SUEZ CANAL

The Suez Canal connected the strategically important Mediterranean and Red Seas, cutting roughly 6,000 miles from the journey for British ships travelling to India as the 100-mile canal removed the necessity to take the existing route around the most southerly point of Africa. This shorter trip meant lower costs in fuel and crew, inevitably reducing the transport costs of the goods carried. The canal presented the unlikely scene of ships bearing flags from all over the world sailing along what seemed, at first glance, to be mere desert.

Ferdinand de Lesseps was the Frenchman responsible for building the Suez Canal. It took him and his team ten years to complete, and they laboured under the motto *Aperire terrram gentibus* ('To open the earth to all mankind').

Egypt is a centre for the joining of three continents, belonging to Africa yet adjacent to the Mediterranean, connecting it to Europe, and accessing the Red Sea with its links to the Arabian Peninsula, East Asia and the Indian Subcontinent.

Ferdinand de Lesseps was not the first to have the vision of a canal connecting them all. Egypt is renowned for its hot, dry desert. The only relief comes from the Nile, which flows north into the Mediterranean, providing along its banks a bed of rich, fertile silt. For many thousands of years, Egyptian pharaohs and farmers have dug canals to direct this fresh water. The ancient Egyptians were not especially interested in linking the Mediterranean and Red Sea as trading routes did not exist at that time between East and West, Orient and Occident, and their relatively fragile boats would not have coped with the rocky reefs and erratic winds of the Red Sea. However, they were keen to establish a waterway with the Red Sea to maximise their trading links with Ethiopia and the Land of Punt (thought to be Somalia today). Gold, ivory and spices were brought from East Africa to ports on the Red Sea, where they would be transported by camel inland to Egypt.

Traders wished for the Nile to be joined to the Red Sea so that they could take their goods to Egyptian cities by a faster and less dangerous route. Since the Nile flows into the Mediterranean, this was to be the first link between the two seas. The call was answered in 1800 BC by Pharaoh Sesostris III. Known as the Canal of the Pharaohs, this waterway was dug by countless prisoners and was just deep enough to allow the shallow vessels of the time to pass through. As well as economic activity, the canal extended political power into remote regions. However, it required constant upkeep to stop the desert sands from filling it. As time passed, and pharaohs rose and fell, the canal was gradually reclaimed by the sands.

In 700 BC, Pharoah Necho II took on the task of recreating the canal. He faced added difficulty as the Red Sea had subsided, divided now from the Bitter Lakes. 120,000 men were said to have died during the digging. Necho himself called a halt to the proceedings after he visited a soothsayer who told him that the canal would be utilised by foreign marauders.

One hundred years later, in 600 BC, Persian ruler Darius the Great built a canal between the Bitter Lakes and the Red Sea so that he could gain better access to his troops in the Persian Gulf. Indeed, when Lesseps was working on the canal almost

2,400 years later, a red granite stone was found with an inscription in an ancient language that was discovered to read:

> I am Darius, the great king, king of kings, king of this wide earth, son of Hystaspes, the Achaemenia. Saith King Darius: 'I am a Persian. From Persia I conquered Egypt. I ordered this canal to be dug from the river called the Nile which flows in Egypt to the sea which goes from Persia. So, this canal was dug, as I commanded, and ships went from Egypt to Persia, according to my desire.'

Later rulers let the canal go to wrack and ruin. Some were fearful that the saltwater of the Red Sea would contaminate the fresh water of the Nile, bringing catastrophic consequences. The occupation of Egypt by the Roman Empire in 30 BC brought new opportunities. Egypt enjoyed its emerging role as a centre for global commerce. The Romans desired foreign goods and set up a new water route from the Nile to modern-day Cairo, which was deep enough for their larger boats to transverse. It was called the Channel of Trajan after their emperor. Exotic animals, herbs and medicines were purchased from the East and taken up the canal and into the Mediterranean Sea.

The Arab invasion of AD 639 brought an end to the Roman Empire in Egypt, and the canal was demolished. This may have been to prevent hostile agents from attacking, or possibly to starve the insurgent inhabitants of Medina and Mecca who relied on imported provisions. The canal was filled in by AD 767 and remained so for over a thousand years.

However, trade continued. Marco Polo, the Venetian merchant and explorer, used an overland course to get to China and India in the late thirteenth century, travelling the Isthmus of Suez – the strip of land between the Mediterranean and the Red Sea – in a very demanding journey.

In 1453, Constantinople was seized by Ottoman Turks. They had control of the land routes to Asia as well as the Red Sea, and they were threatening to outsiders. European merchants were in danger. The Portuguese explorer Vasco da Gama, in 1498, managed to get to India by traversing around the Cape of Good Hope in South Africa, keeping away from the Turks. The overland route to India no longer mattered. Britain and Portugal, with coasts facing the Atlantic Ocean, were inadvertently lucky. Indeed, both had a good supply of reliable ships and excellent navigators. Once a highway, the Mediterranean Sea had become a cul de sac.

Although the journey took longer, there was only one point at which to load and unload, saving time and money. Spices could be sold at a fifth of the price in Lisbon, Portugal's capital, compared to Venice, where they were imported overland. Portugal therefore gained control of many markets, and even the Venetians were forced to buy from them.

The Venetians subsequently urged the Egyptians to build a canal linking the Mediterranean and Red Seas. The Venetians were unable to find the money for the project and the Egyptians were unwilling to finance an undertaking that would help foreigners. By the beginning of the 1500s, ideas had moved on to the notion of a canal that linked the two seas via the Nile. The Turks were especially behind the idea, as were the French who saw Marseilles, situated on the Mediterranean, suffering from the change in trade routes. Pope Sixtus V was also eager about the canal, envisaging his missionaries to India enjoying a quicker journey.

Money was a problem. No one had the funds to create a new canal. Even excavating the old one was not possible. By the end of the 1700s, politics in Egypt had become so confused that it did not take long for the French army, led by Napoleon Bonaparte, to seize control. His prime minister, Talleyrand, gave him precise instructions:

The Army of the East shall take possession of Egypt. The Commander-in-Chief ... shall have the Isthmus of Suez cut through and he shall take the necessary steps to assure the free and exclusive possession of the Red Sea to the French Republic.

Napoleon was overjoyed at this order. He hoped to thwart Britain's power by the building of this canal, noting in his journal that it would be 'as fatal as the discovery of the Cape of Good Hope was to the Genoese and Venetians in the sixteenth century'. He acknowledged that Egypt was a vital region in world affairs, commenting that 'who is master of Egypt is master of India', adding, 'Really to destroy England we must get possession of Egypt.' This view was felt keenly by the British, who were worried about their trade with India. One merchant warned, 'France, in possession of Egypt, would possess the master-key to all the trading stations of the earth ... she might make it the emporium of the world ... and England would hold her possessions in India at the mercy of France.'

Jacques-Marie Lepere was the engineer commissioned to survey the isthmus and cost the digging of the canal. Lepere wrongly advised Napoleon Bonaparte that the level of the Red Sea was 32 feet above that of the Mediterranean. Therefore, if a canal was made between the two there would be extensive flooding unless locks were put in place. His recommendation was to excavate the Canal of the Pharaohs, which he thought would take 10,000 men four years to do at a cost of £1.5 million.

However, his evaluations were incorrect, and events would conspire to prevent the canal's progress anyway. The French lost a battle against the British and Turks in Alexandria in 1801 and had to retreat home. The French influence in Egypt, for now, was finished. But such were the politics of the day that the canal's future was still far from over.

An Englishman named Thomas Waghorn, who had served in the British Army in India, subsequently went on to work for the East India Company in Cairo. In 1835, he trailblazed an overland route for the movement of mail from Alexandria to Suez. By the late 1830s, he had managed to transport and store huge amounts of coal at the port of Suez, so that steamers as well as large ships could be docked there. Waghorn also conveyed passengers along this route, which involved a desert trek of twenty-four hours. Donkeys and camels were used in the blistering heat, and constant negotiations had to be made with the Bedouins for safe passage. Waghorn's route was generally disregarded by all except for the Frenchman Ferdinand de Lesseps, who would eventually build the canal, and who went so far as to commission and erect a statue of Waghorn, declaring at the canal's opening ceremony in 1869, 'He opened the route. We followed.' De Lesseps went on,

> When English navigators pass by this monument to Waghorn, erected by the French, they will remember the intimate alliance which should always exist between the two nations that have been placed at the head of world civilisation, not in order to ravage the world, but in order to enlighten it and bring it peace.

Though Egypt at that time was still under the control of the Ottoman Empire of Turkey, in the early nineteenth century, a young man called Muhammad Ali was appointed Viceroy, ruling the country from 1805 to 1849 like an independent nation. Muhammad knew that both Britain and France wanted to take control of Egypt, and he was determined to modernise the country and make it strong enough to break free of Turkish rule. He had been friends with Mathieu de Lesseps, Ferdinand's father, for some time, and even though the latter was a French diplomat they enjoyed each other's company enormously. Ferdinand grew up in Pisa, where he had private tutors and enjoyed a relatively luxurious life. The family moved to Paris when he was ten, in the wake of the Hundred Days War and Napoleon Bonaparte's defeat at Waterloo in 1815, which led to the second restoration of King Louis XVIII Ferdinand in July that year.

As a young man, Ferdinand was keen to join the foreign service, starting in 1825 at Lisbon, where France had a key interest. The French were attempting to annex Spain, which was under the rule of a Bourbon monarch. Ferdinand enjoyed mixing with the ruling class in Iberia in his position as France's spokesman, but also as cousin to the Comtesse de Montijo, who was carrying the baby that would grow up to become the Empress of France and a prime mover in the canal's future. After a time, de Lesseps was moved to North Africa. In 1832, at the age of twenty-six, he became vice-consul in Cairo.

He had his work cut out for him. His boss, Mimaut, left his post in 1835, and de Lesseps was promoted to counsel. At this time Muhammad Ali was looking to increase his power, taking Constantinople from the fragile Ottoman Empire, even though Egypt was under the rule of the Ottoman Sultan. In 1833, Egypt sent troops, led by Muhammad Ali's oldest son, Ibrahim. His army fought the Ottoman Empire at the Battle of Nezib on 24 June 1839, where they successfully defeated the Ottomans, giving them control of the whole of Syria. The Turkish Sultan, Mahmud II, died prior to hearing the result of the battle; his successor was just sixteen years old. That autumn, the Egyptians headed towards Constantinople where their success seemed certain. In gaining the city they would take control of the entire eastern Mediterranean. Muhammad Ali relied on backing from France and Spain to finish the job, but the Ottoman Empire's allies, including Russia, Britain and Austria, wanted him out of Syria. The Sultan was actually willing to relinquish Syria to the Egyptians, but his allies were not.

The Ottomans' allies sent troops to back them up, forcing an Egyptian retreat. The British were spurred to act because they relied on the Ottoman Empire to ensure the sea routes between the Black Sea and the Mediterranean remained open and free of Russian control. Although France had been weakened by the Napoleonic Wars, their signing of the Congress of Vienna in 1814–15 still allowed them to fight abroad. In fact, most European nations at this point entered conflict through representation and the eastern Mediterranean was a key battleground.

The Eastern Question, as it was known, was the conflict between the great powers of Europe over the Ottoman Empire, which ranged from south-east Europe, inclusive of the Balkans, to the Arab Peninsula. The Ottoman Empire, at that time, was referred to as the Sick Man of Europe, and was viewed to be erratic and fragmenting, riddled with corruption, misgovernance and barbarity towards ethnic minorities, especially Christians.

De Lesseps knew that in the regions of the eastern Mediterranean, France and Britain were in stiff opposition. The French campaign in Egypt and Syria from 1798 to 1801, in which Napoleon Bonaparte had brought 4,000 soldiers as well as scientists and artists, was an attempt to weaken British trade routes to India while promoting French endeavours. The two campaigns of 1833 and 1839 had heightened suspicion between the two nations, and this feeling was extended towards the construction of the Suez Canal. In years to come, de Lesseps was met with enormous scepticism in Britain when he spoke of his ideas.

However, fate took its course. On his way to take up his post in Egypt, a fellow traveller on de Lesseps' ship was struck by what was thought to be cholera. As a result, everyone on the boat had to be quarantined when they reached Egypt until the nature of the illness could be understood. The Consul General of France provided a box of books to Ferdinand to read whilst he passed his time in quarantine. Among them was a ten-volume set describing Napoleon's ventures in Egypt, incorporating Lepere's work on the canal. The project captivated Ferdinand, and so began a lifelong fascination.

When he finally reached his post, Muhammad Ali took the young Frenchman under his wing, telling him that, 'it was your father who made me what I am'. Ferdinand was a keen athlete who soon took to swimming and riding the Arab ponies for hours across the desert. Muhammad's youngest son, Muhammad Said, was rather a chubby boy, weighing almost 200 lbs at the age of thirteen. His father put him on a strict exercise regime, asking Ferdinand to teach him to ride. This Ferdinand did, and Said was also expected to take part in other routines such as rowing, rope climbing and skipping. He would often make his way to Ferdinand's kitchen, ravenous and ready to drop. Ferdinand fed him macaroni. The compassion that he showed the boy would bring unexpected benefits in the future.

Another character with whom Ferdinand came into contact was the Frenchman Prosper Barthelemy Enfantin, leader of the Saint-Simonians. Their mission was to further world peace, not through politics but by virtuous deeds which enhanced friendship between nations. Enfantin wanted to join East and West, spiritually and culturally, through the construction of canals at Suez and at Panama. These ideas became Enfantin's *raison d'etre*. But the group lost credibility as their strong religious fervour saw them anoint Enfantin as the new Messiah. They held extensive, doctrinal worship sessions, processing through the streets confessing their sins, wearing priestly tunics of various shades of blue.

Muhammad Ali wanted to know whether the Suez Canal was possible and asked de Lesseps if he would liaise with Enfantin, who assured de Lesseps that Lepere's survey was faulty and that the canal was possible. However, Muhammad Ali was in truth more interested in building up his armies than in building a canal, and Enfantin returned to France, albeit still eager to pursue his vision. He set up an international body of engineers, operating under the name of the *Société d'Etudes pour le Canal de Suez*, or the Society for the Study of the Suez Canal. They were concerned primarily with the engineering aspects of the canal, and with raising funds from European capitalists; they knew from the beginning that the canal would have to be an international effort. From Great Britain, Enfantin was advised by Robert Stephenson, who also examined the railway from Cairo to Alexandria. The society went to Egypt in 1847 to survey the possible canal route. Muhammad Ali was highly mistrustful of them as he thought that European leaders were behind the scheme, eager to sabotage Egyptian independence.

The study group continued their work throughout 1847. De Lesseps may have been kept abreast – one letter shows him to be keen – but he was busy with his diplomatic responsibilities. Enfantin and his society, meanwhile, were making progress. They sent French engineer Linant de Bellefonds to re-measure the sea levels and he found out that they were in fact even, meaning a difficult lock structure would not be needed. Satisfied, Enfantin got in contact with both the King of France, Louis-Philippe, and

Muhammad Ali, and brought them onside. The canal project was set to go ahead, using French engineers and with Enfantin heading the scheme. But then disaster struck. Louis-Philippe was forcibly removed from power in 1848, and soon after the eighty-year-old Muhammad Ali died. This was the end of the line for Enfantin and his society.

Muhammad was succeeded by his nephew Abbas, a heartless and volatile man who had been orphaned as a child. He took it upon himself to revoke Muhammad Ali's forward-thinking schemes, seeing them as unwise and hazardous. He brought home the Egyptian scholars who were studying in Europe and closed educational establishments devoted to science. Many of Muhammad Ali's modernising ministers were sentenced to death. Abbas denied access to merchants from Europe, effectively sealing Egypt off from international trade.

De Lesseps was at this point working in Rome, trying to sort out conflicting demands from the many armies that occupied the city. The Austrian forces were against the French, and between them was the recently created Roman Republic, which wanted the Pope to forgo his political authority. The Austrians stood by the Pope and wanted to destroy the Roman Republic. The French army backed their new president, Louis-Napoleon (nephew of Napoleon Bonaparte, and later Napoleon III), whose political motives were unclear. However, the French general Oudinot, desired the use of force against both the Roman Republic and the Austrians. De Lesseps was in the thick of things. His rather obtuse orders were 'to deliver the States of the Church from the anarchy which prevails in them, and to ensure that the re-establishment of a regular power is not in the future declared, not to say imperilled, by reactionary fury'.

It was a very difficult situation to manage. De Lesseps had arrived in Rome in early May and spent hours in talks with General Oudinot and Republican leaders, travelling back and forth between the two. Everybody was tense, but a temporary result was gained later that month which appeased the Republicans while also allowing the French forces to stay in the city. De Lesseps requested a meeting with the French foreign minister to report on his work. He found, on his return to Paris his role had been occupied by another man, Alexis de Tocqueville, who gave permission for General Oudinot to conquer the city. De Lesseps was held responsible for going beyond his orders and for achieving an invalid agreement. Tocqueville did not blame him but did not save him either, and de Lesseps was thus deprived of his post. However, that was not the end of it. He was taken before the Council of State and accused of compromising the renown and distinction of the national army and thereby sabotaging the honour of France. He was then formally chastised, effectively ending his career as a diplomat. He retired at the age of forty-four to live in the French countryside.

Abbas Pasha did not last long. His death sparked the arrival of a new Viceroy, and it was de Lesseps who was celebrating this time. His old friend Muhammad Said had risen to power. De Lesseps wrote immediately to the new Viceroy, who was now going by the name Said Pasha. He got a swift reply with an invitation to meet at Alexandria, which he received with joy.

De Lesseps departed from France at the end of October 1854. History cannot tell whether or not he visited Lyon, where he could have stopped in to see François Arlès-Dufour, a main player in Enfantin's Canal Study Group, gaining papers that he could use to persuade Said to take up the endeavour.

Said Pasha had gained yet more weight. He was quite an unpredictable character. Weak-willed and easily led, while mainly calm he was liable to sudden fits of depression or anger. De Lesseps played the submissive role, which won him favour, and he was granted the use of a palace decorated lavishly with silks, marble and solid silver hand-basins. A team of servants was provided to fulfil his every need. De Lesseps was undoubtedly the golden boy.

The stream of Europeans who had converged on Egypt during the reign of Muhammad Ali before disappearing in the barren years of Abbas now returned with dreams of making their fortunes. Said was gullible, and easily taken in by newfangled inventions and grand schemes. One of his many palaces overlooked a vast parade ground where he enjoyed watching his army perform training exercises. The ground here was extremely dusty during the summer months, and one enterprising European suggested he cover it with iron. This he subsequently did, at a vast cost, and the dust was gone; instead the burning heat of the sun caused the soldiers' feet to be scorched, even through their boots. Said was quickly developing a reputation as a man who could be easily swindled.

De Lesseps was still nervous about mentioning the canal, especially when Said Pasha declared that he had no interest in it. De Lesseps knew that he had to play his hand carefully to get his point across, and he got his moment when Said Pasha invited him to watch some army drilling in the desert between Alexandria and Cairo. They spent several days getting there, with de Lesseps eager to press his case. He witnessed a rainbow one morning, extending from east to west, and his heart leapt – here was a symbol of all he wanted to achieve. The other tipping point, according to de Lesseps, occurred when he leapt on an Arabian horse, the sun rising behind him, and showed his prowess in jumping. Said Pasha was impressed and took him to one side, and so they began their conversation about the Suez Canal.

De Lesseps assured Said Pasha that the Canal would guarantee the safety of Egypt for years to come, especially from a European occupation, and it would exhibit that the country 'still has the capacity to be a potent force in world affairs, and is still capable of adding a brilliant page to the history of world civilisation'. De Lesseps was certain that the technical aspects of the work, although difficult, were achievable, and it was really the political and economic challenges that needed to be addressed. He appealed to the Viceroy's vanity:

A canal would set Said apart from other rulers. It would transform him from the governor of an Ottoman province into a potentate admired throughout the world and immortalised as a man who dared to do what others said was impossible. 'The names of the Egyptian sovereigns who erected the

Pyramids, those useless monuments of human pride, will be ignored. The name of the prince who will have opened the grand canal through Suez will be blessed century after century for posterity.'

'The Canal,' de Lesseps continued, 'would secure the passage of pilgrims to the holy sites of Mecca and Medina and make the ruler of Egypt a protector of the faithful. The canal would connect Europe to the lands of India, China, Japan, and Australia, and place Egypt at the centre of world trade. It would enrich Egypt and bring it more revenue than cash crops ever could.'

Said Pasha gave him his attention, enjoying the heady mix of eminence and fortune, and having heard de Lesseps plans he replied, 'I am satisfied. I accept your plan. For the rest of the journey we will concern ourselves with the means of carrying it out. You may regard the matter as settled, and trust to me.'

Within two weeks an agreement had been signed, with de Lesseps granted exclusive rights to form an international company to build the canal on a ninety-nine-year lease from the launch of the canal, after which it would be run by the Egyptians. In the meantime, 75 per cent of profits would go to shareholders, 10 per cent to the canal founders and 15 per cent to the Egyptian administration. Said Pasha was to grant the land required for the project, which at the time was unproductive. The company would have it for ten years after the opening of the canal, after which time they would reimburse the government with a levy. This land, however, once the freshwater canal had been built, was to become some of the most high-yielding, and therefore expensive, in the world. It was stipulated that work could not proceed until the leader of the Ottoman Empire, the Turkish Sultan, had agreed.

There was, nonetheless, opposition to the canal. Britain was the world's leading maritime nation and had robust trading links with India and the Far East. If the canal was to be built, other European states could easily usurp that position. Enfantin also made a complaint, saying that he had a prior agreement with Muhammad Ali. De Lesseps contended this with the agreement he had just signed with Said Pasha. De Lesseps did not want to work with Enfantin, who was furious upon learning the fact. De Lesseps took two engineers to survey the Isthmus, declaring that 'there lived not even a fly in this appalling desert'.

As De Lesseps began to consider the work ahead of him, the task of bringing thousands of workmen to the region looked increasingly impossible. Their first visit alone needed the support of fifty camels, half of them bearing water, and this was only a three-week reconnaissance. Nevertheless, they collected information on the depth and length of the proposed canal, as well as materials that would be needed. He also set up the International Scientific Commission, a body of scientists and engineers from Europe and Russia who would give advice on the project.

Another challenge was persuading the Turkish Sultan to approve the work, and by extension the British. The Ottoman Empire at that time was failing, but European countries were happy to support it for various reasons, one being that they did not want the threat of a Russian advance into their continent. Lord Stratford de Redcliffe, the British Ambassador to Turkey, was reputed to have immense power over the Sultan, and de Lesseps realised that to get the Sultan's approval he would also need de Redcliffe's. De Redcliffe advised the Sultan to defer a decision about the canal as Turkey was fighting the Crimean War against Russia with both Britain and France as allies. This was an uneasy relationship for all concerned as historically Britain and France had never seen eye to eye. The Sultan did not want to upset the peace between these two nations.

However, disaster was to strike at the heart of the British Empire when in March 1857 the Indian Rebellion occurred. With many British lives being lost the British government deployed new troops to the Subcontinent, and the only way they could get there was by the four-month trek around the Cape of Good Hope. In the end, the government had to implore Said Pasha for permission to send their troops overland across the Isthmus of Suez. The British press were furious with their government's weakness, the London *Daily News* declaring that 'nothing could be a more complete avowal of the utility of M. de Lesseps' scheme and the action of the Government is the implicit condemnation of Lord Palmerston and Lord Stratford de Redcliffe'.

The British Prime Minister, Lord Palmerston was extremely negative about the canal, thinking that it would be regulated by the French and Egyptians who would deny unrestricted access to the British. The British press envisaged that the canal would be liable to sandstorms that would bury it. De Lesseps appealed to the British people, telling them the canal would create a shorter and more secure route to India and the Far East. However, try as he might, he could never get around Lord Palmerston.

Palmerston was a commanding presence in British politics for nearly half a century. With untamed hair and exuberant sideburns, he presented quite a bad-tempered figure. He was Foreign Secretary for a large part of the 1830s and 1840s, and he created the template of the inflexible, extremely patriotic British oppressor. He was Prime Minister from 1855 to 1858, and once more from 1859 to 1865, when he died. He believed in England and the righteousness of the British Empire and Crown. He was not one to form allies with other countries, and he posed a significant threat to the creation of the canal.

Palmerston was reluctant to allow the British people greater democracy by extending the vote and liked the traditional setup of aristocratic rule. However, novels by Charles Dickens and the social movements of Victorian times were becoming ever more popular, leading to increased democracy being demanded by the growing middle class. De Lesseps was aware of these factors, and he believed by appealing to the British people he could force the government's hand. He did not realise the depth of feeling and suspicion in the country against the French. He met with Palmerston on a number of occasions, with the two men at loggerheads; de Lesseps believed the canal would bring global advantages, while Palmerston was convinced it would have the opposite effect. De Lesseps began to believe Palmerston had a persecution complex, while the British Ambassador to Paris, Lord Cowley, sought to intervene by explaining how the French for some time had deliberately undermined the building of the British railway in Egypt. Lord Cowley wrote to the British Foreign Secretary, explaining, 'It would be a suicidal act on the part of England to assent to the construction of this canal.'

The British were unsure of Napoleon III's support for the canal. They were reluctant to disapprove publicly in case de Lesseps and Napoleon were in cahoots, so they took the path of reluctant aversion, as did the Ottomans, probably for the same reason. De Lesseps carried on his publicity campaign, telling the British people the same story that he had told in Cairo and Constantinople. The canal would cut the journey to the Arabian Sea from around 11,000 miles, as it stood with the Cape of Good Hope, to a mere 6,000 miles. The canal would be neutral, and his vision was to have capitalists from all over the world investing in it.

It was a difficult path to travel. De Lesseps went across Europe selling his ideas. It was politically sensitive, of course, since the Ottomans wanted to keep the peace with the British and French and to retain their slipping influence in Egypt. Said approved of the canal, but not at the cost of losing peace with Britain and Turkey. The French, too, were behind the canal, but not if it brought aggravation from Britain. De Lesseps had to work with all these groups and somehow find funding for the canal. He mustered all the diplomatic prowess that he possessed, using different perspectives with multiple factions in order to achieve his goal. Said funded most of the advertising campaign. An independent report in 1856, carried out by a team of internationally acclaimed engineers, came back with the wording, 'The execution is easy; success is assured, and the results will be immense for the commerce of the world.'

Palmerston met de Lesseps one more time during the Frenchman's tour of Great Britain in 1857. De Lesseps had spent three months drumming up support among the British public and had one up on Palmerston when he met him – the people were behind him. Parliament debated furiously the whole question of the canal in July 1857. Many wanted Lord Stratford to ease pressure on the Turkish Sultan. The rebellion in India that spring created an intense pro-canal sentiment among those who had previously been indecisive. However, Palmerston continued to undermine de Lesseps, calling the canal an impossible scheme that would ruin Britain's influence in Egypt and Turkey

The famed engineer and MP Robert Stephenson was called in again, this time by Palmerston, to fight the project. One of his speeches in the House of Commons infuriated de Lesseps so much that he challenged Stephenson to a duel. Stephenson backed down, but de Lesseps still did not understand why Stephenson would dispute the engineering feats which experts across the world considered quite feasible. Realising that the British government would never come round, de Lesseps gave up on them.

At this time Said Pasha granted a new concession, confirming the straight path of the canal and the creation of a freshwater canal from the Nile to Lake Timsah. He added that ports were to be erected at both Timsah and Pelusium. The agreement specified that 80 per cent of the workforce would be made up by Egyptian nationals using the corveé system which had existed in Egypt for centuries. Each man was to be paid two to three piastres daily, with the Canal Company – founded by De Lesseps in 1858 – providing drinking water, food, tents and travel allowances. The tendency towards 'slave' labour was part of the Egyptian culture and no one really thought anything of it.

Enfantin, meanwhile, was furious and went to the French emperor, saying that the Saint-Simonians had been working on the project for twenty years before de Lesseps came along. He reminded Napoleon of the diplomatic scandal surrounding de Lesseps, and warned that the man had no experience of leading a construction project bigger than the work he had done on his rural manor. Enfantin was keen to regain control of the project, feeling that in de Lesseps' hands it would come to nothing. He pressed the emperor to support his prior claim, saying that it would control the spread of Russia in the East as the Crimean War had done in Europe. Napoleon ignored Enfantin's protestations for some time, at last brushing him off with a caustic note. Enfantin's venom had got the better of him, while no one could claim de Lesseps was anything but sincere.

Said did a lot to promote the idea of the canal with the Ottoman Sultan. He travelled to Constantinople in the autumn of 1856, primarily to win the Sultan's approval. This he did not get. Said held the purse-strings over the canal and could halt the project at any time, and de Lesseps was keenly aware of this fact. They travelled together to Sudan in November 1856, with de Lesseps trying to ignite in Said his earlier enthusiasm. Said was finding it difficult to balance the strain that the canal was creating between himself and both Britain and the Sultan, however. De Lesseps found his political position was of far greater concern than the dream that he was so keen on pursuing.

Britain's overseas interests had been steadily growing for hundreds of years, but it was not until the mid-nineteenth century that they gained supremacy over the remainder of the European continent. The Great Exhibition at Crystal Palace in London in 1851 cemented their position, with Queen Victoria declaring that the 'hand of God was visible in the many wonders of human artifice assembled there. Britain was blessed among all nations, and the future beckoned with unimagined wonders.'

Palmerston and his circle stood for Britain's solitary dominion, and they got it. In the 1850s, the British navy controlled the seas. Their military ruled India, Australia, New Zealand, Hong Kong and parts of Canada, but their real power was demarcated by trade and industry, which, under the leadership of Palmerston, was flourishing. Yearly exports increased between 1850 and 1870 from £80 million to £240 million. This was intensified by the coal from Wales and Newcastle, powering steelmaking plants for new railways. The factories in the Midlands used raw materials from India, the United States and Egypt to produce clothing which in turn was sold with large financial gains to other parts of the world, such as China, mainland Europe, Argentina and India. Countries like the United States were able to source their own raw materials, but Britain's natural wealth was limited and it needed to trade in order to industrialise. This led to a combative, self-determining foreign policy, with their powerful navy providing security. In essence, ruling the waves was absolutely vital to the continued success of the empire. Palmerston was extremely suspicious of the Suez Canal project, as was Lord Stratford de Redcliffe. The latter was regarded with awe in Constantinople and was so dreaded that he was called Sultan Stratford.

Meanwhile, the International Scientific Commission had been working on the plan to take the direct route from the ancient setting of Pelusium on the Mediterranean due south through the parched Bitter Lakes onto Lake Timsah and to Suez and the Red Sea. Their surveys planned for a canal 103 miles in length, extending 26 feet down, with a width of almost 80 feet at the base, allowing transport of the biggest vessels of the day. They estimated the project would take six years to complete using men and spades. The cost of the work would be in the region of 200 million francs.

There were three ridges of rock and sand throughout the length of the proposed route. El-Guisr Ridge, on the Mediterranean side of Lake Timsah, was 10 miles long and stood 30 feet above sea level. Serapeum Ridge was located between Lake Timsah and the Bitter Lakes and was of a comparable size. It was the great mass of Shallufa which would cause the biggest problems. It was located 13 miles to the north of Suez, south of the Bitter Lakes. The correct machinery was needed to excavate these rock faces and Shallufa continued to be an obstacle right up until the opening ceremony.

De Lesseps was becoming vexed about the finances of the project. He eventually went to the wealthy bankers of the Rothschild family. He met Baron James Rothschild, who appreciated the canal scheme and offered monetary assistance. A delighted de Lesseps got up to leave when he realised something and turned back to Rothschild.

'What do you want in exchange?'

'Good Lord! It's easy to see that you are not a businessman! It will be the usual 5 per cent.'

'5 per cent!' said de Lesseps, 'but on two hundred million francs that is ten millions! Do you propose to take ten millions of money away from my shareholders in payment for the use of your dingy corridors? No, thank you. You may keep your bank. We will make issue without you. I will hire some suitable office for which I will pay twelve hundred francs a month, and it will do me quite well.'

'You will not succeed.'

'We shall see.'

'The bankers wished to lay down the law to me,' de Lesseps reported later, 'but I would not give in. I decided to do my business quite alone and go direct to the public.'

De Lesseps went ahead with his original idea of floating the stock to the public, so that the company would be made up of thousands of shareholders instead of being in the hands of the affluent few. He was eager to get on, not wanting to wait for the Sultan's approval. Now that he had a plan of action, he went ahead with raising money, rationalising that once he had sufficient shareholders it would be harder for the Sultan to dismiss the enterprise. So, in October 1858, de Lesseps made public the Universal Company of the Maritime Canal of Suez, and from 5 November started selling shares in the company at 500 francs each. Investors were to gain 5 percent interest on their shares and a percentage of profits when the canal was finished. The work itself was costed at 160 million francs, but allowing for interest payable to shareholders, the amount rose to 200 million francs. This figure would be recovered by freight charges on the traffic which would be diverted from the route round the Cape of Good Hope.

De Lesseps met with Queen Victoria in London and had a long chat with Prince Albert in his study. Prince George, Duke of Cambridge was said to have given his unreserved support. The Frenchman was the guest of honour at the Society of London where they gave him rounds of applause. However, the British government continued to pour scorn on the scheme. When the shares came to be sold, Russia, the United States and Austria bought none. The 20,000 shares that had been kept in hand for Britain were unsold also. France, however, came up trumps, buying 201,111 shares when only 80,000 had been allocated for them. They were purchased by 21,000 French people and represented more than half the shares available.

People from all walks of life invested: judges, mechanics, engineers, priests, teachers, doctors, bankers, soldiers, clerks, lawyers – mainly from Paris, but also from the industrial hubs of Lyon and Bordeaux, as well as the port of Marseille, which would benefit enormously from the canal. Many were buoyed on by the fascination in Egypt brought about by Bonaparte's attempted conquest there in 1798 and his subsequent publication of the cultural heritage of the pharaohs and other aspects of Egyptian life. Shares were also bought in Spain, Denmark, Portugal and Prussia. Said Pasha showed a lot of support by purchasing most of the shares reserved for the Ottoman Empire as well as taking most of the unsold shares. He became the biggest shareholder with 177,642 to his name. It was a huge financial undertaking, but he had already promised to do just this. De Lesseps' expectation that Said stay true to his word caused a degree of friction between the two men, although they both remained inspired by a united vision. Palmerston's fear that the canal would be a French and Egyptian enterprise came true. Having mulled it over, Said Pasha confessed himself delighted:

> This scheme is a child of my own begetting – a scheme I have looked upon as the one great act of my life … I cannot but consider the junction of the two seas as an undertaking which will immortalise him under whose auspices it is carried out.

The idea of a joint stock company was still fairly alien at that time. Traditionally, the government or the monarch covered the cost for large public works, or, at a more local level, the aristocracy or civic authorities. It was only when the railroads were built twenty years previously that the public had been encouraged to lend their financial support. An international company was also a very new idea, and while many countries knew they would benefit from the canal it was nevertheless seen to be a French endeavour.

As time passed Said Pasha grew apprehensive, not just about his great outlay but also the reaction of the Sultan to news of the canal's slow progress. He lost his nerve when de Lesseps told him that the company was in a position to start work. Said Pasha could not bear the thought of conflict between the Turks, the British and the Canal Company and decided he was unwilling to give his permission to begin, telling a British diplomat that 'not a sod shall be turned ... until the Sultan's sanction shall have been obtained'.

The international community were unsure of de Lesseps' relationship with Emperor Napoleon III. De Lesseps' cousin, now Empress Eugénie, was Napoleon's wife. Therefore, politicians did not want to upset the emperor by dismissing de Lesseps. His position was unclear, but the fact that he was operating without an official title made foreign administrators believe that Napoleon was holding his cards close to his chest. Napoleon came to power in 1848 when he called a national referendum after King Louis-Phillipe had been overthrown that February. Some 7.8 million people voted him in, compared to 300,000 who voted against him. The Second Republic was born, and in Napoleon's eyes peace in France equated to world order.

However, that peace was shattered later in the year when the Italian Felice Orsini tried to assassinate Napoleon on 14 January 1858. He did not succeed, but he killed twelve people on their way to the opera. The attempt had repercussions for diplomatic relations between France and Britain as the plotters were based in England. The French insisted that the remaining conspirators were handed over. Palmerston refused, as he did not like violation of British jurisdiction, creating murmurs of war as the French were furious. This led to Palmerston's enemies succeeding in removing him from office for the following year.

The British were terrified of France gaining power at their expense. An international academy had deemed that the canal was possible and that at 200 million francs it was cost effective since over 12 billion francs had been spent in the last twenty years building railways in Europe. In June 1858, the House of Commons held a debate about whether to use their influence on the Sultan to push for the cancellation of the canal project. Their fears of a new French empire won out.

The British were also putting pressure on Said Pasha, and de Lesseps was getting quite fed up. On 25 April 1859, he decided to go ahead with the canal anyway. With his team of engineers and 150 local workers at the historical Mediterranean site of Pelusium, he hoisted an Egyptian flag and commanded the men take their pickaxes and make their first blow. With words from de Lesseps honouring Said Pasha, the men commenced their work, removing small quantities of sand with their spades. When all was said and done, the project would see the removal of over 100 million cubic feet of sand, mud, sludge and rock.

News of their actions caused fury among the British, who commanded the Turkish Sultan to get Said Pasha to stop at once. De Lesseps continued unabated. He was taking a great risk, not only of losing the support of Said Pasha but of provoking British military aggression. De Lesseps was not going to give in and responded by writing a firm letter to Said Pasha, reminding him of his financial involvement and the fact that if he were to back out he would have to reimburse the other shareholders. This troubled both Said Pasha and the British, as if the Viceroy needed cash he might relinquish the British venture that had eaten up so much time and money: the railroad linking the Suez to Alexandria, opened in 1858. The British did not want this to happen, so they quietened down.

The strain of these international tensions took a toll on the relationship between Said Pasha and de Lesseps. The latter had claimed rather forcefully that the company was now an irreversible legal enterprise and by law must be allowed to carry out its mission. He was single minded, 'sometimes discouraged but always defiant, planning, persuading, publicising, calculating and manoeuvring … a man of action above all, he was profoundly practical'. He realised that although Britain had backed down, it was only a temporary reprieve. Said Pasha was not in the best of health. It was important for de Lesseps to move on with the canal;

> The work must not stop for an instant, whatever the obstacles placed in the way. The strength of his *fait accompli* lay in the Canal's concrete progress, and though its momentum might have to slow down it must at no cost be arrested. Were this to happen, all might be lost.

The work itself was fairly elementary – or so they thought. The line between the seas was reasonably straight and there were five desert lakes along the route. Four of these were dry; it was only Lake Menzaleh that contained water. The soil to the north was dirt and sand, whereas the terrain to the south was clay and gravel. Little thought was being given to the problems posed by the rock formations of Shallufa, the Serapeum and El-Guisr. None of the European engineers had dealt with these conditions before.

That was only half the story. Labour needed to be brought in, and these men needed to be fed, watered and kept free from disease. Work started at the Mediterranean, constructing a harbour. To begin with, huts for 3,000 men were needed. The harbour was required so that dredging machines and seaborne cranes could be transported to the digging sites, although during the beginning phases of the canal, pre-industrial labour would be used, and it was only as the canal progressed that mechanical methods were implemented. As the site was windy a breakwater was needed, and it was thought that putting rocks in the sea would detract from the power of the waves. There were, however, no local materials to achieve this. There was only sand. The nearest quarry was all the way over in Mex, near Alexandria, and it would cost a great deal to get them to the site.

The engineers came up with a site-specific solution: they made the boulders that they needed. They created a mould, which was filled with a blend of lime mortar, saltwater and sand, and left to dry. The block was then ready to be submerged in the waterfront. Over 30,000 rocks were required to construct the harbour, which de Lesseps named after the Viceroy: Port Said. These boulders would be worn down by saltwater and only have a shelf life of around thirty years, but the engineers were unperturbed; by that time the canal would be generating money and real rocks could take their place.

Special dredgers and floating cranes had been designed by the Suez engineers to work on the harbour, but most of the work was done in the time-honoured way, by hand. The Egyptian workforce was to become a subject of intense political scrutiny. Egypt had been called the breadbasket of the Mediterranean since Roman times due to the unpredictable rise and fall of the Nile, which sometimes flooded towns but always provided irrigation, creating the incredibly fertile soil of the area.

Food was provided for the workers by the town of Damietta, which was nearly 20 miles from Port Said. Water was more problematic. The wells that were dug produced such saline water that not even the camels would drink it. The local fishermen used ancient distillation methods that again proved quite foul. Eventually, at quite a cost, water had to be transported from the Nile, which lay to the west. Over 3,000 camels and donkeys took this cargo to barges that lay on Lake Menzaleh, and from here it was taken to the construction site.

Said Pasha was concerned about the British constantly berating the project. Regional administrators obstructed the water supply, with violent assaults on de Lesseps' men when they dared ask about its progress. Water became a key issue and could have compromised the whole project. De Lesseps countered this by digging the Sweet Water Canal, straight from the Nile to the construction site, so that they were able to utilise the freshwater stream to transport tools, food and other requirements, as well as water to the thousands of labourers. Men were working in temperatures of over 120 degrees Fahrenheit (49 degrees Celsius), so this was a necessary development. This freshwater canal went along the route of the ancient Pharaoh's

Canal, and it took over two years to build with offshoots going north and south to service different parts of the construction project. Water from rainfall alone was not enough to provide for the extensive workforce that amounted to thousands of men. At Port Said de Lesseps also constructed distilling equipment that boiled saltwater. Only a finite amount of water could be distilled each day, and security guards were placed to safeguard the instruments.

Some within the Egyptian court were furious that de Lesseps had started work without permission. But he stood his ground, telling Said, 'The adversaries of the Suez Canal are the enemies of Egypt and of the Viceregal dynasty. Your Highness is not at liberty to dismiss the sacred engagements, contracted in full view of the civilised world, and I am not free to delay ... their fulfilment.' Napoleon, however, decided to lend his support to the canal, even at the expense of British criticism, as the French public who had invested in it were vital to his popularity.

Port Said was completed in December 1860. The desert had been transformed into a town containing shops, houses, a 50-foot lighthouse and a jetty reaching 6,000 feet into the sea. 120 Europeans lived there, as well as 250 Arabs. It was quite an achievement.

Next up was Lake Menzaleh, which was really more of a swamp. De Lesseps and his team travelled by camel and mule to survey it. The British loved to scorn the idea that the 4- to 5-foot-deep lake could possibly be excavated to allow big ships to traverse it. They had a point; the bottom of the lake was full of what was known as slob. This was a build-up of sediments from the Nile, with a huge proportion of water and sulphur. It could not be heaved up and placed on the side of the excavation as it simply slid back in. Work ground to a halt as none of the men had worked with this slob before. But local fishermen, who were used to the geology of the area, were brought in to help. Armed with the knowledge of centuries, they were able to grasp the slob and squeeze it in their arms. Water would be lost, and they then took it on their backs to the shore where it would dry in the heat of the sun. In all, 14 million cubic feet of material was excavated using this method, with the walls created at the side of the lake drying into a rock-hard structure 6 feet tall.

The Sweet Water Canal was completed in November 1862 and was able to supply the increasing number of workers. The branch connecting it with the Suez was also ready, enabling de Lesseps to dispatch men to the region. About 50 miles of the canal had now been completed, from Port Said south to Lake Menzaleh and on to Lake Timsah. The canal was deep, with strong walls, and water flowed freely throughout it, only being contained at Lake Timsah by a slender bank of mud. The workers would allow some water through to the area where they were digging if they needed to soften the soil. De Lesseps wanted to mark these achievements with an event in which he would allow the waters dammed up by the ridge of mud at Lake Timsah to flow freely with those of the Mediterranean Sea. Said Pasha was not able to attend due to heart problems, but de Lesseps employed an Egyptian band to play their national anthem, after which he said a prayer for Christians as well as for Muslims. Then his workers broke through the bank of mud. Loud shouts were heard as the water came pouring through. The world press showered their praise on de Lesseps.

The British authorities were furious with his achievements and became even more committed to stopping his work. They took the tactic of lambasting de Lesseps for his facilitation of what they called 'slave labour', entreating both Said Pasha and de Lesseps to emancipate these men. Said Pasha, ever dithering, decided to please the British by restricting the quantity of local workers. The Viceroy was in financial difficulties. In 1862, he had to pay his first 15 million francs to the Canal Company as part of his instalment plan on his shares. This was a sizable amount in relation to the taxes paid to his government, and he looked to Europe for a loan. Palmerston attacked, saying that Said had been swindled by de Lesseps, that he would destroy the Egyptian economy and the French bankers would take over Egypt to regain their credit.

Said Pasha died on 18 January 1863 at the tender age of forty-one. This made things very difficult for de Lesseps, who despite their difficulties had always considered him to be a close confidant. He expressed a deep sadness, 'not because of my enterprise, for which I maintain the greatest faith … but for the cruel separation from a faithful friend who, for 25 years, gave me so much evidences of affection and confidence'.

Said's nephew Ismail Pasha took over as Viceroy. He was thirty-three years old and had been educated in Paris at the Ecole d'Etat Major. Said had sent Ismail as a special envoy to the Pope, as well as representing his interests with the Sultan and Napoleon III. Ismail was despatched to Sudan to quell an uprising in 1861, which he successfully accomplished. De Lesseps was nervous about the future of the canal. Said Pasha had left debts of over 400 million francs, with high rates on interest accruing on these national loans all the time. Ismail was impetuous and loved luxury, wanting to transform Cairo into the Paris of the Middle East. To do so, he built a splendid opera house and numerous luxury palaces, widened the streets and created parks. Ismail restructured the post office, education system and customs, invigorating trade. He borrowed money from European banks at escalating interest rates without having any plans about how to repay his creditors. His illusions of grandeur convinced the Ottoman Sultan to amplify his position, granting him the superior title of Khedive of Egypt rather than Viceroy. He was keen to gratify the British, but he did claim to be 'more of a Canalist than Monsieur de Lesseps'.

The British were piling on the pressure about slave labour, which would ultimately cause quite severe headaches for de Lesseps. The contract signed by Said Pasha and de Lesseps in 1854 had set out that Said Pasha would provide four-fifths of the labour. These men would receive 2½ to 3 piastres payment for each day. They would be given housing and medical attention if need be, and the Canal Company would also pay for their travelling costs. Doctors were provided to guard against cholera outbreaks, which were common in Egypt, and also to take care of the nutritional requirements of the men. Egyptian officials and Canal Company men would be on hand in each settlement to keep the peace. However, Britain still made noises about the unsuitability of using slave labour in a modern world.

These Egyptian 'slaves' were known as the fellahin, who worked under a structure called the corvée. By law, they were expected to drop everything if the government required labour for a building project; they were not paid. They had been working on the canal project since its inauguration and the British decided to influence not just the Khedive but also the Turkish Sultan to stop this labour. Slavery was a hot topic. The Civil War in the United States at that time would result in the abolition of slavery there. The Tsar in Russia was under attack pressure to end serfdom. Attitudes in Britain, where the slave trade ended in 1807, and France (1848) were strongly against slavery. The Sultan wanted to appease the British as he needed their help in keeping Russian forces from entering Turkey. De Lesseps was finding it nearly impossible to recruit paid workers. He managed to get a few hundred Egyptians and a small number of Syrians and Palestinians on his payroll, but that was not nearly enough. He was embarrassed that his state-of-the art concept was being achieved by slave labour. These were men who were usually coerced to work temporarily on irrigation projects on the Nile Delta, work that benefited them as well as the ruling class. Not only that, but due to the American Civil War cotton was becoming more difficult to obtain. Egypt saw rocketing prices in the sale of their long-staple cotton, which multiplied their annual income fivefold to around £25 million. Palmerston got on the bandwagon, addressing Parliament on the subject:

It is very much to be regretted, in the interest of both England and France that when both countries are much in need of cotton, 30,000 or 40,000 people who might be usefully employed in the cultivation of cotton in Egypt are occupied in digging a Canal through a sandy desert and making two harbours in deep mud and shallow water. I should hope that so useless an occupation will soon be put to an end.

Nubar Nubarian, a European-educated Egyptian diplomat, continued the debate. He worked out that 20,000 fellahin arrived at the canal each month while the same number left. The total Egyptian population stood at 4 million, and 720,000 of them were employed on the canal across the year. Egypt could have gained an extra 36 million francs each year if these men had worked in the cotton fields. De Lesseps, in reply, continued to use his well-worn argument that the canal would unite Muslims and Christians in the formation of a better world.

With so much money from cotton available, Ismail started work on Egyptian infrastructure, including a railway, irrigation and land reclamation, sugar-refining works and upgrades to the cities of Cairo and Alexandria. But the money was not guaranteed to last forever. By 1865, peace had been declared in America and they were able to start trading cotton again. Egypt lost a valuable source of income, but Ismail did not stop spending. He borrowed heavily in his quest to bring Egypt in line with European development, which was important in his eyes, for his status and for protection.

The British meddling angered de Lesseps. He argued that he was a good employer. He actually paid the fellahin, and did not expect them to dig more than one cubic metre a day even though it was perfectly possible to dig three times that. He also held the British to account for their double standards: they had used the fellahin to build their train line in Egypt, which had caused a great number of deaths. Conditions for miners and factory workers in Britain and France were as reprehensible as those for the fellahin. The UK had pressganged men into the navy up to the defeat of Napoleon Bonaparte in 1814 and the Treaty of Fontainbleau. He questioned why the British had not made any comment on the use of slaves in the United States or Russia. But the Sultan was resolute. By April 1863, he ordered that the maximum number of fellahin employed by the Canal Company would be decreased to 6,000. Land that Said Pasha had given the Canal Company in their contract had to be returned. The Sultan warned that if these conditions were not met he may have to resort to force. On 1 January 1863, Abraham Lincoln had delivered the Emancipation Proclamation, granting slaves in the south of the United States their freedom, leading to increased pressure that Ismail do the same.

The Sultan's move infuriated de Lesseps. The Canal Company did not have the funds to take on new workers. He resorted to contacting the Emperor of France, Napoleon III, to see if he would agree to a meeting between him and Ismail. Ismail wanted to stay on good terms with the French, and de Lesseps, knowing that the French people had invested heavily in him, wanted to get the canal finished.

Napoleon was in favour of the canal. He stipulated that the Egyptian government would have to reimburse the Canal Company to the tune of 38 million francs to make up for the loss of the free labour of the corvée. Napoleon also decreed that the fertile areas to the side of the Sweet Water Canal, which were initially pledged to the Canal Company, gaining a revenue counteracting the cost of excavating it, were to be compensated to the tune of 30 million francs. This put the Egyptian government in a very tricky position. Their final bill to pay to the Canal Company was 84 million francs, with 10 million coming from a bulk payment to the company and 6 million for navigation permits.

This was the same as the income and expenditure for the entire country per year, but Ismail was not downhearted. He had the greatest number of shares in the company and stood to gain 15 per cent of profits when the canal was eventually utilised. Buoyed up by cotton prices, he took out a loan from the Oppenheim brothers in Germany to the tune of 100 million francs, with a 15 per cent dividend for the brothers. This was his first step towards bankruptcy and the ownership of his country by European powers.

De Lesseps managed to get the Sultan to sign the contract too, to his great exhilaration, as he was fed up with his project being used as a diplomatic haggling point between the Turks, Egyptians and British. It led de Lesseps to gleefully announce that the canal 'no longer was a hope, but an incontestable fact'.

De Lesseps was grateful for the emperor's decision and turned his attention back to the excavating of the canal. Word had spread and new recruits swarmed in from England, Belgium, Syria, France,

Italy and all over Africa, keen to be involved in such a momentous project. It was running behind time, and these new workers still did not add up to the numbers the company had employed using the corvée. Machines became increasingly important in getting the work done; there were as many as sixty of them at work at any point in time. By diverting water from the Sweet Water Canal, the dredgers were able to sit on the water while they worked. Men and machines were sent to the Red Sea to begin work there, advancing north. The canal became a hive of enthusiastic young engineers all looking for experience on the most ambitious project in the world. By the mid-1860s, over 8,000 Europeans had flocked to the site, among them accountants, clerks, shopkeepers, doctors, surveyors, carpenters and many more – all of whom wished to gain the skills, high rates of pay and contacts that the project offered.

The influence of the Egyptian government decreased as the fellahin were no longer used. The Canal Company began to operate in a seemingly self-governing way, building new towns along the route of the canal, numbering over 20,000 people, and administering the day-to-day activities. De Lesseps was becoming a very powerful man and was losing the need to gain authorisation from Ismail. His actions in the long run would be extremely dangerous for the Egyptian nation, leading to the banishment of the Khedive, the bankruptcy of the government and occupation of the country by foreign powers.

In April 1865, when the American Civil War was drawing to a close, and Egypt was beginning to be viewed as a developing power, de Lesseps took some United States representatives on a tour of the canal. One of these men, our old friend Cyrus Field, was the leader of the New York Chamber, and had been busy working with George Elliot on the laying of the first Transatlantic telegraph cable. He was in awe of the project, telling de Lesseps,

You have undertaken to divide two continents for the profit of all the commercial nations of the world; I fervently pray that you will soon see complete success, and that this work will rest as a monument as durable as the Pyramids because of your energy and talent.

Then, in May 1865, something terrible happened. A cholera outbreak began, initiated by pilgrims travelling from Mecca towards Alexandria. During the following two months, hundreds of workers died as it spread throughout the entire canal. Death occurred a day or so after symptoms first appeared, and very little was known about the disease at that time. De Lesseps had spent a great deal of money on a wide-ranging and costly health service for his workers. The fact that the illness was spread through contaminated water was only just beginning to gain ground.

There was a mass exodus. Everyone was trying to get home. Thousands flocked to Port Said, although many hundreds met their end at Ismailia. De Lesseps did his best to restore order and stop the panic, realising that people moving around so much would only spread the epidemic. The outbreak lasted throughout the summer months, and although there were quarantines, 1,500 Egyptian and Arabs died, as well as a few thousand Europeans. De Lesseps lost his own grandson to the illness. It was not until nineteen years later in 1884 that Robert Koch was able to identify the causative agent of cholera, spending much of his time in Egypt in order to do so. The death of de Lesseps' great adversary, Lord Palmerston, in that same year at least reduced the pressure on him.

By the autumn of 1865, de Lesseps realised that he was facing another financial crisis. There had been too many delays, and there was no hope of getting the work finished on time. He needed an extra cash injection. At his most optimistic, he could hope to get the work finished by the summer of 1868, but that would require the machines to operate without a hitch and no unseen errors to occur. The company was adroit at hiding the amount of money it had, in comparison to what it was spending, but shareholder meetings were open to accountability, and de Lesseps knew that at some point in the coming year more money would have to be made available unless the shareholders were left with an incomplete project.

Rather than devalue the price of existing shares, de Lesseps decided to raise the 100 million francs required through a public loan. He was now in a position where the canal had become a symbol of national French pride, and he was worried that a crisis in funding would reflect badly on his country. He decided to raise the issue during the Universal Exposition of Art and Industry, held in Paris throughout the spring and summer of 1867. Napoleon was hoping to rival England, who at that time led the way in industry, aping their Great Exhibition at Crystal Palace in 1851. Little did he know it, but time was running out for Napoleon and his Second Empire.

Over 11 million visitors enjoyed the exposition, among them the rulers of Europe and the rest of the world, including the Ottoman Sultan and the Khedive. The Ottoman Empire was still powerful, controlling what would later be known as Albania, Bosnia, Bulgaria, Armenia, Egypt, Kuwait, Iraq, Jordan, Syria, Saudi Arabia, Israel, Lebanon, Serbia, Libya, Tunisia and Turkey. The exposition was a place where people could exhibit the latest trends in mechanical engineering and perhaps even engage a patron. De Lesseps made the most of the opportunity to present a Suez display, planning on announcing his need for further patronage once visitors had been sufficiently wowed. It worked, and some of the money came in by September. De Lesseps said that he wanted to operate a lottery to raise the rest of the funds, but the French government was unimpressed, wanting more transparency for the project. However, the lottery was eventually accepted as a means to raise the funds, with a Parisian newspaper writing, 'Public attention is now fixated on the piercing of the Suez Canal, and it is impossible not to consider it a political enterprise as well as an industrial and commercial endeavour.'

De Lesseps grew in fame as Napoleon withered. He was called 'the most remarkable man of this era', and he lapped up the praise and attention that was lavished on him. Khedive Ismail also received a great deal of interest and respect from the Paris crowds. He thoroughly enjoyed his trip to Europe, and returned to Egypt ever more determined to spend, spend, spend and create a country that could rival what he had seen. The cotton industry in Egypt was not so lucrative now that the Civil War in the United States had ended, so he increased the output of sugar by building expensive refineries. However, this market did not perform as he had hoped and he generated further debt. Ismail took out more loans that he simply could not afford, but he was leaning heavily on productivity as a source of future income. He was the biggest shareholder in the canal and was eligible for 15 percent of the profits once it started operating, as well as a percentage from customs duties. Realising the canal's worth to his purse, he now considered de Lesseps a great confederate. But de Lesseps had bigger fish to fry. With the backing of so many powerful people he turned his attention back to completing the canal.

In May 1867, the company had set its fees. They based their calculations on the 10 million tons of goods that were transported from Europe, Asia and the United States each year, hoping that 5 million would make use of the canal. Their initial figure of 20–25 francs per ton was thought too high, and it was reduced to 10 francs per ton, with a further 10 francs for each traveller. In their eyes, this would create an annual income of 60 million francs. However, they grossly overestimated how long it would take for companies to transform their habits.

By the end of 1867 only half the canal had been dug. The route between Port Said and Lake Timsah was complete, and although used by barges and smaller ships, it was still not ready for bigger craft. To the south of Lake Timsah and Ismailia lay the 6-mile Serapeum ridge, which had yet to be worked and which prevented the flow of water southwards to the Bitter Lakes. These lakes, thankfully, had little need of quarrying, but they led to the craggy Shallufa ridge and the Suez. A rockface of 5 miles stood in the way of the teams. Using tons of explosives, they slowly managed to make a foot or a yard's headway at a time, with the sound of great blasts resonating through the air. There was concern, as time drew on, for the final flooding of the canal, especially the Bitter Lakes, which were 8 miles wide in areas. No one knew how the pressure of the water was going to affect the canal – even whether it would destroy it.

De Lesseps took the bold step of scheduling a grand launch party for 17 November 1869. He had announced a completion date of October 1869, so it was a race against time. Money was running out, yet again, and he needed to get the work done. Thousands flooded to the canal site in search of work. In 1865 the canal was home to 20,000 people, but by 1868 this had doubled in number, with individuals from all over bringing their own businesses with them. Ismail, having won some of his independence from the Ottomans, found that his fate was linked to the canal, and with it the autonomy of European powers. It was integral to de Lesseps' plan that the canal was a neutral shipping route. De Lesseps also stated that ships would not pay customs duties on goods that were not meant for Egypt; that is, they would not pay if they were just passing through. The fine detail had never been worked out on this score. Ismail was furious and took the matter to arbitration, and it was noted internationally that de Lesseps meant his scheme to be more than a business enterprise. Maybe he was modelling it on the East India Company, which had its own military and sought to rule the provinces it dealt with under the auspices of the British monarchy. The Khedive, however, anxious to get the work done, did provide the Canal Company with the funds necessary between 1868 and 1869.

As an interesting aside, the young Frédéric-Auguste Bartholdi had spent many years in Egypt and the Holy Land studying the work of those ancient cultures. He proposed to de Lesseps a plan for a sculpture of a huge female Egyptian fellahin holding a flame which would act as a lighthouse for the ships entering Port Said. Both the Khedive and de Lesseps rejected the project on financial grounds. Bartholdi settled in the United States in 1886, and his sculpture was presented in New York Harbour, as part of the 100-year anniversary of the Declaration of Independence. He named it *Liberty Enlightening the World*, and we know it today as the Statue of Liberty. De Lesseps was among those to attend the unveiling.

By August 1868, almost 27 million tons – a third of the total – still had to be excavated with only thirteen months to go. The jetties in Port Said were completed at the beginning of 1869. The machines that had been in use since 1863 were working seventeen hours a day, but the company was worried enough in 1869 to hire thousands more labourers. They worked on the 10-mile route from the Shallufa ridge to the Suez, using their hands to remove the sand.

Work carried on throughout the night, with resinous wood providing light in overhanging braziers. Machines were dredging the length of the canal, with the towns at Port Said, Ismailia and Suez now being finalised. The Prince of Wales (the future Edward VII) went on a trip to the Suez in spring of 1869 and was deeply impressed by what he saw. The Bitter Lakes were being filled. He grumbled that, in disagreeing with the canal, Lord Palmerston 'had been guilty of a lamentable lack of foresight'.

By mid-August 1869 the Bitter Lakes were almost full with sea water from the Mediterranean and the Shallufa ridge was no more. The remaining few miles to the Suez had been excavated using unskilled labour. The last obstacle was the dike built to confine the Red Sea from travelling north. The dyke was removed on 15 August 1969, with a small ceremony, connecting the waters. The initial deadline had passed by a year ago, but de Lesseps was over the moon. The opening festivities were upon him, and they would in themselves be extravagant and costly. Unfortunately, there was a small hiccup: over 100 yards of rock was still protruding in the Shallufa region, just below the surface. Dredgers were not able to remove it, and an emergency gunpowder supply had to be brought to the area just as the flotilla of boats were about to depart from Port Said on the inaugural journey.

Meanwhile, the Khedive was facing quite serious conflict with the Ottomans in Istanbul. Tensions were rising, and both were concerned that armed conflict might follow. De Lesseps was trying to woo the shipping industries of the world, but he was finding it hard to drum up much interest. People were very sceptical about the safety of the canal. Would the canal simply be choked with silt? Could the strong currents of the Red Sea be made safe? De Lesseps decided that his launch party was the best way to assuage these anxieties. He worked with the Khedive to secure a million-franc budget and sent out 1,000 invitations for people to have a free trip down the canal as part of the ceremony. The Khedive footed the bill for the most part, as the Canal Company was strapped for cash.

The grand opening of the canal was still set for 17 November 1869, with Empress Eugénie as guest of honour. A flotilla was planned in the harbour of Port Said, with the ships processing through the canal over a three-day period. The cities lining the canal would be the focus of celebrations, culminating in a grand party in Cairo.

Eugénie had been entranced by Egypt and the canal at the Universal Exposition and made good friends with the Khedive while he was there. A yearning of escapism reigned in mid-nineteenth-century Europe, particularly a desire for the romanticism of the past. Napoleon had never really been a good husband to Eugénie, and his empire was beginning to disintegrate. She longed to see ancient Egypt and to take the limelight, without her husband, who could not be coerced to go. She planned on sailing her yacht, *L'Aigle*, first to Venice, then Constantinople and Alexandria, fitting in a two-week sightseeing trip in Egypt before sailing to Port Said for 16 November. Following her example, talk of the opening ceremony was all the rage in Paris. Fashion and theatre reflected this new trend. The newspapers were full of stories on the Egyptian climate and how the canal would affect business in Marseilles. Package tours were arranged, and de Lesseps invited artists and writers on a free tour of the canal. The English, meanwhile, were rueful. Their papers reported on the nation's failure to take a large part, bitterly lamenting the scientists and diplomats who had reported that it was an impossible venture.

The Sultan declined an invitation to the ceremony; tensions between him and the Khedive meant he was in no mood to celebrate Ismail's achievements. However, many of the European heads did accept, giving de Lesseps the headache of having to plan the order of ships entering the canal. August saw the canal's annual shareholder meeting, at which de Lesseps had to disclose the cost of the canal so far: 404 million francs, nearly twice the 200 million francs he had initially foreseen. More money – tens of millions of francs – were needed to complete the canal and to pay the required dividends to the shareholders. What is more, he was trying to hide from everybody that the canal would need more work even if it opened on time.

Although Ismail had been made Khedive of Egypt in 1867, rather than Viceroy, which gave him a little more power, he was still under the jurisdiction of the Turkish Sultan and the Ottoman Empire. What his greater title *did* mean, and what he had paid substantial bribes for, was that his first-born son would inherit the leadership on his death rather than it passing to the eldest male relative. He was looking forward to the canal opening as it would be an international occasion and a method of increasing his status, leading to greater autonomy. The newly created town of Ismailia on the banks of Lake Timsah in what had once been barren desert had been named after him. He established a home there among de Lesseps' family and a great number of the canal's officials and engineers. Ismail was keen for the opening ceremonies to be as grand as possible, so while de Lesseps informed the newspapers, he set off on a journey around Europe to personally invite the monarchs and heads of state to the event. The ceremonies would last for two weeks. All expenses were to be paid, including transport costs to and from Egypt, with hotels, food and entertainments being all part of the package.

Most recipients were excited about the invitation, either being able to attend in person or nominating an appointed official. The United States was not so forthcoming. The president could not attend, and as Congress was not sitting nobody could be found to take the lengthy trip. The celebrations were due to start in Cairo, and many flocked to the city. Among them were monarchs, politicians, businessmen, financiers and lawyers as well as journalists, artists, historians and soldiers. Arab leaders brought their tribes and their animals, setting up camp outside the city.

A thousand servants were employed to tend to the visitors, with five hundred chefs imported from Europe. No expense was spared: new linen for beds, damask curtains and luxury Egyptian cotton towels coloured to match the national flag of each guest. The Khedive went so far as to build a new opera house for the opening ceremonies, more elaborate than those found in Europe.

Empress Eugénie had backed her cousin de Lesseps throughout the whole project. She was to open the canal, and Ismail was particularly keen that she enjoy her stay. She had shown an affection for the Pyramids, so the Khedive went to great lengths to light up the whole of the Great Pyramid and built a luxury villa at its base. Egyptian ministers were becoming increasingly concerned about the expense of the opening ceremony. However, Ismail was not to be moved. He created a 10-mile boulevard between the Pyramids and Cairo to replace the rough road. Although it was built by the fellahin in six weeks, who were whipped for their troubles, there was not a murmur of disapproval from international quarters.

Excitement was high in the Khedive's camp. However, not everything was going according to plan. Fireworks worth thousands of dollars had been hurriedly deposited in a timber works at Port Said. Needless to say, one of them caught fire and rapidly set off all the others. Some 2,000 troops had to be deployed to quench the flames and avert a catastrophe.

De Lesseps had his own concerns. He was anxious about the enormous boulders which had just been found in some areas of the canal. He was desperate, as it was a possibility that some of the ships might hit these rocks as they moved through the canal in the grand procession. His fears were confirmed when they sent the *Latif,* an Egyptian warship, on a test voyage through the canal at 11 p.m. on the eve of the opening ceremony. By 1 a.m. the following morning, the frigate had run aground on some mudflats close to Kantara and was totally obstructing the route. Efforts to tow it away had failed and both de Lesseps and the Khedive were beside themselves. The Khedive finally gave the order to blow up the ship, at which point de Lesseps hugged him in relief. As it happened, over 1,000 men ventured into the mud to free the boat.

Eugénie arrived in Alexandria on 22 October but was not overly impressed by the city and headed directly for Cairo. There she was treated to entertainments and delicious food that reminded her of medieval Spain, which pleased Ismail as that had been his intention. She watched belly dances and the Sufi's rapturous swirling dervishes, and was taken on a barge on a tour down the Nile, remarking on the ancient pyramids and the near-naked villagers and their dismal dwellings.

The first day of November saw the inaugurating opera at the Cairo Opera House. Ismail had commissioned Verdi, the most famous composer in Europe, to write a special opera for the occasion. Verdi was not terribly enthusiastic, but for a fee of 150,000 francs he eventually came up with *Aïda*, a love story set in the time of Ramesses III. It was not ready until December, so a performance of Verdi's *Rigoletto* accompanied the inauguration instead.

On 16 November, boats thronged in the 500-acre waters of Port Said, with ships spilling out into the Mediterranean to watch the event. There was a noticeable scarcity of British vessels. The air was billowing with flags and bunting, and excitement levels were high. The Khedive's yacht, the *Mahroussah*, and Empress Eugénie's yacht, the 300-foot *L'Aigle*, were ready to start the procession. Thousands of people lined the banks and that afternoon a religious thanksgiving was held. People from all over the world took part, in multicultural costume, with diplomats and royalty festooned with gems adorning swords and uniforms. De Lesseps stood out in his sombre suit. The spiritual ceremony started with an imam, followed by an archbishop from Jerusalem and finally the empress's priest, who spoke of the unity the canal would bring to the two faiths.

The desert had been replaced with the thriving towns of Port Said, Ismailia and Suez. The Place de Lesseps at Port Said was flowing with freshwater fountains. Ismailia had extensive gardens and colourful villas and the rockface of Shallufa, standing sentinel over the canal, was planted with exotic flowers. The freshwater canal was teeming with pleasure boats, and small towns had developed along its length, which was full of fish. Grass grew extensively throughout the route and trees such as orange and acacia lined its path.

The ships waiting in the harbour were eager to move on, and fired their cannons in commemoration, causing the onlookers to cough and splutter into their handkerchiefs with all the smoke. The *L'Aigle* was the first ship to go down the canal with de Lesseps and Empress Eugénie on board. Forty-six other ships carrying the international guests followed on. As she made her way, cheers of jubilation could be heard seemingly in every language of the world.

The ships travelled at eight-minute interludes, allowing a kilometre between each vessel. Empress Eugénie was said to be extremely nervous. Any mishap would undermine the French national interest, and she spent much of the journey praying, unable to sit with the others calmly at the table. Her yacht was extremely cumbersome with its 60-foot beam, and the bottom of the entrance to the canal had only been cleared to a small margin. At one point, the captain of her yacht was constantly commanding halting the boat, while Eugénie paced the decks, until finally they were clear of the area.

The procession sailed on to Ismailia, where they landed at Lake Timsah for further entertainments. Picnics and camel races were held throughout the day, and there was to be a huge feast later. There were reportedly 40,000 people in the town that night, and the atmosphere was jovial, with over 10,000 fellahin and Bedouins gathered around the town to share in the celebrations. Canal Company officials welcomed people into their homes. There were riding and rifle displays, and music and entertainers from all around the world. The Khedive financed the whole affair. Nobody paid a penny.

18 November 1869 was the occasion of a ball, held at the Khedive's palace. It was a crush as people turned up regardless of whether they had actually been invited. Pickpockets were at work, and although there was ample food the queues were so long that many decided to look elsewhere. A firework display over the boats on the canal precipitated the dancing.

The next day was also Ferdinand de Lesseps' sixty-fourth birthday. The procession of boats continued down towards the Red Sea. People, horses and camels followed along the banks. Some, who had had too much to drink, did not leave until the following day, but the boats of the various countries vied to get the best place. Although they had to take care in navigating the Serapeum, once they arrived at the Bitter Lakes they scattered and competed against each other. All that was left was to pass through the tight waterway of the Shallufa ridge, which had only just been created. From there to Suez was a simple journey, and the town itself, although not as decorative as Ismailia, was full of people all cheering their approach. The ships had all passed successfully through the canal. The venture was a triumph.

Empress Eugénie's yacht was the first to enter the canal, followed by the British P&O liner *Delta*, but Eugénie's vessel was not the first to complete the trip to the Red Sea. That honour went to Sir George Nares, in his British vessel HMS *Newport*. He had spent the previous night getting his ship to the front of the queue in pitch blackness so that no ship was able to overtake him on the narrow finishing straight. Nares was officially reproached for this humiliation of the French, but behind closed doors he was highly praised and given the rank of captain.

De Lesseps' fifteen-year venture was accomplished. He had joined the Mediterranean and the Red Sea. On 25 November he remarried to a young woman named Hélène Autard de Bragard at a small gathering in Ismailia. She was to deliver him twelve children.

De Lesseps received great acclaim for his efforts, even from the British. Queen Victoria awarded him the highly distinguished Grand Cross of the Star of India. The Prince of Wales wrote to him that 'Great Britain will never forget that it is to you alone that we owe the success of this great achievement'. This was high praise from a nation which had done everything possible to destroy de Lesseps' dreams. However, Britain did allude to this in an oration by the Lord Mayor of London, who said, 'Our engineers made a mistake – M. de Lesseps was right, and the Suez Canal is a living fact.'

However, the following year bought financial troubles as only half the expected 1,000 ships used the canal. De Lesseps hoped the tolls would bring in 8 million francs, but they only realised 4 million francs. The 10 francs a ton charged on cargo seemed by many to be too expensive, and the carriage of the ships in the first year amounted to just over 400,000 rather than the 5 million tons anticipated. Britain made up three-quarters of the total. The Canal Company faced ruin. The share price fell considerably, and dividends went unpaid. The canal had cost 500 million francs to build. De Lesseps, having once been a hero, was now an object of ridicule. However, the land along the canal was still prospering and de Lesseps, with his vast quantity of canal shares, remained wealthy. Some even said that Ismailia would take over as Egypt's capital city. De Lesseps was awarded the cross of the Legion of Honour, or Légion d'honneur, by his cousin the empress. But things were about to change.

In the summer of 1870, Napoleon III announced his intention to do battle with Prussia. His empire was promptly crushed on 1 September that year as the Prussian army defeated the French at the Battle of Sedan, where Napoleon was captured. The emperor and his wife, Eugénie, were banished, and the Third Republic was created in France. The French took some time to get back on their feet as they did not have the mercantile or naval resources to utilise the canal a great deal. Napoleon died in Britain in 1873, with Eugénie living there until 1920. She would become close friends with Queen Victoria and remembered the canal's opening as the pinnacle of her reign.

Many ships were staying away from the canal due to the Red Sea's lack of wind, which their sails needed in order to navigate. Other vessels were too large for the canal, the depth of which still needed to be improved. A number of ships had already been grounded.

Shareholders were eager to see their investment come to fruition. The lack of traffic combined with the more than doubling of construction costs created uncertainty and the selling price of shares plummeted by 50 per cent. The years 1870 and 1871 were difficult for de Lesseps and the Canal Company. A British newspaper of 30 December 1870 observed:

> The Suez Canal has of late dropped out of sight, but it may be interesting to state that the traffic receipts are now averaging £20,000 per month. M. de Lesseps, the soul of the undertaking, is shut up in Paris.

It was not until steam power became more prevalent in the shipping world that his luck changed. Towards the end of 1872, this new form of power was taking precedence and the canal at last was seeing a profit.

The Third Republic in France was led by men of industry and power. They praised de Lesseps' work as one of France's proudest accomplishments. De Lesseps lessened the tariffs on freight, and soon the Canal Company was turning a profit. He was now asked to advise on other endeavours, including the excavation of a canal in Panama joining the Atlantic and the Pacific. This was a project that de Lesseps was very keen to begin, with the help of his son Charles.

In the meantime, Ismail was hurting financially. His debt continued from the disastrous expenditure of 1869. In 1873 he borrowed £30 million, an amount that was impossible to repay as he spent £10 million just to reimburse interest. He was playing a very dangerous game, modernising his country and amplifying his nation's income on eye-watering credit. By the middle of the 1870s, his debts had increased to almost £100 million with a very high rate of interest. Lenders in Europe no longer believed he could repay his debts and stopped giving him money. Indeed, they thought his country was about to go bankrupt. By November 1875, when his latest interest payments were due, Ismail had nothing left to give and no one was prepared to help. He had just one resource left.

First Ismail met the French banker Edouard Dervieu, who had worked for him before. Dervieu approached both the Société Générale and the Anglo-Egyptian Bank, who had previously loaned money to the Khedive. Ismail offered them his shares. They spoke to Henry Oppenheim, a member of the Oppenheim Syndicate, which had formerly been responsible for lending to the Khedive at very high interest rates. He in turn spoke to Baron Lionel Rothschild and Frederick Greenwood, the editor of the *Pall Mall Gazette*, who notified the British government.

Benjamin Disraeli, the British Prime Minister since 1868, had already shown an interest in buying the Khedive's shares in the summer of 1876, and on hearing the news he was quick to respond. He was worried about the French gaining complete control of the canal, as it was proving useful to British shipping and important in maintaining the British Empire's control in Africa and south-east Asia, where the French were also doing well.

In pounds sterling, the canal had cost the Egyptian government £16,075,119 so far in construction costs, equating to US$86 million, or 429 million francs. The government had borrowed, for this and other projects, £35,437,474 payable at interest rates of 12.4 per cent for the state loan of 1873 and 26.9 per cent for the 1866 railways loan. Ismail had used all Egypt's resources as security for these loans, and by 1875 the only money he was receiving came from his shares in the Canal Company. In order to settle these differences, he took to taxing the Egyptian people. The vast majority did not have sufficient funds and this burden drove them further into penury. As early as 1867 Lady Duff Gordon, a visitor to Egypt, had written:

> I cannot describe the misery here now – every day some new tax. Every beast, camel, cow, sheep, donkey, and horse is made to pay. The fellahin can no longer eat bread; they are living on barley meal mixed with water, and raw green stuff. The people in Upper Egypt are running away by wholesale utterly unable to pay the new taxes and do the work exacted. Even here the beating for the year's taxes is awful.

Benjamin Disraeli and Edward Henry Stanley, Lord Derby, his Foreign Secretary, were alerted to the offer of the shares on 15 November 1875. With Britain as the most extensive user of the canal, Disraeli had always desired that the nation take a bigger role in the running of the canal and a greater share of its revenue.

Disraeli had travelled extensively in Palestine and Egypt in the 1840s and penned a handful of novels about life there. One of these was *Tancred*, in which the hero finds the meaning of life not from the progressive industrialisation of the West but from the joining of the cultures of East and West.

Disraeli's Conservative Party held a majority in the House of Commons, and he was on very good terms with Queen Victoria. He was in a position whereby he could make the decision on the purchase of the Suez Canal shares on his own, with the rational assumption that Cabinet would back him up. Queen Victoria, he imagined, would also be happy for this matter to proceed. She would see it as an expansion of her empire, and did not want to have to listen to the upstart Prussian Otto von Bismarck talking about whether Britain was pre-eminent enough to control an empire.

Lord Derby responded to word of the Khedive's willingness to sell by telegraphing General Stanton, the Consul-General in Egypt, asking him to find out more information. He also communicated with Lord Lyons, the British Ambassador to Paris, saying, 'The acquisition would be a bad one financially and the affair might involve us in disagreeable correspondence both with France and the Porte [the Ottoman government].' The reply from General Stanton indicated that the Khedive would rather his shares were sold to England than elsewhere.

A total of £4 million would be needed to buy the shares. Wanting to get the deal done quickly, Disraeli acted without the permission of Parliament, which was in recess. He asked his great friend Lionel Rothschild for assistance. Lionel had previously brokered a deal at the conclusion of the Franco-German War in 1871, supporting the structure of foreign exchanges. De Lesseps had of course gone to James Rothschild when he was setting up the Canal Company and had been shocked by his interest charges, even when they were the going rate. Disraeli was offering only a little more than the French, who were struggling to raise the money in time. Disraeli wrote to the queen immediately:

Confidential
2, Whitehall Gardens, S.W.
Nov 1875
Mr. Disraeli with his humble duty to Yr. Majesty
The Khedive, on the eve of bankruptcy, appears desirous of parting with his shares in the Suez Canal, and has communicated, confidentially with General Stanton. There is a French company in negotiation with His Highness, but they purpose only to make an advance with complicated stipulations.

'Tis an affair of millions; about four, at least, but would give the possessor an immense, not to say preponderating influence in the management of the canal.

It is vital to Her Majesty's authority and power at this critical moment, that the Canal should belong to England, and I was so decided and absolute with Lord Derby on this... that he ultimately adopted my views and bought the matter before Cabinet yesterday. The Cabinet was unanimous in their decision, that the interest of the Khedive should, if possible, be obtained, and we telegraphed accordingly.

Last night, there was another telegram from General Stanton, (not in reply) which indicated some new difficulties, but the Cabinet meets again today (at two o'clock:) and we shall consider them.

The Khedive now says, that it is absolutely necessary that he should have between three and four millions sterling by the 30th of this month!

Scarcely breathing time! But the thing must be done.

Mr. Disraeli perceives, that, in his hurry, he has not expressed himself according to etiquette. Your Majesty will be graciously pleased to pardon him! There is no time to rewrite it. The messenger for Balmoral is waiting. He thought Yr Majesty should know all this, and could not write last night, as fresh intelligence was hourly expected.

On 19 November 1875, Disraeli contacted the queen again:

The pecuniary embarrassments of the Khedive appear to be very serious, and it is doubtful, whether a financial catastrophe can be avoided. The business is difficult, but it is as important, as difficult, and must not be relinquished – We received telegrams from General Stanton, who had personally seen the Khedive, and we also returned telegrams.

The Khedive voluntarily pledged himself, that, whatever happened, Yr Majesty's Government [should] have the refusal of his interest in the canal. All that can be done, now, is to keep the business well in hand.

Disraeli still faced stiff opposition from his cabinet, including Lord Derby, but when they gathered on 23 November his private secretary, Montagu Corry, was positioned outside the door while they talked. After the debate, Disraeli opened the door and said a simple 'Yes' to Corry, who hastily departed for New Court to meet Baron Lionel where they exchanged the following words.

'How much?'

'Four million pounds.'

'When?'

'Tomorrow.'

'What is your security?'

'The British Government.'

'You shall have it.'

Baron Rothschild was able to offer the required money with only a day's notice. The Khedive was in no position to disagree; his debts were huge and he had run Egypt dry. As he had used his shares as security to borrow money elsewhere, the British Government would not legally be allowed to draw dividends from them until 1894, but he was desperate to sell to get hold of the £4 million urgently.

Baron Lionel telegraphed the Khedive's finance minister, Ismaïl Seddick, to let him know that £2 million would be available to the Egyptian government on 1 December 1875, with another £1 million on the 15th and the final £1 million in early January 1876. The cabinet still had to get the consent of Parliament, but they did not see that as being a big problem. The shares were to be delivered to General Stanton, HM Consul-General in Egypt.

Disraeli contacted Queen Victoria on 24 November 1875:

Mr. Disraeli with his humble duty to Yr Majesty.

It is settled: you have it, Madam. The French Government had been outgeneraled. They tried too much, offering loans at an usurious rate, and with conditions wh: would have virtually given them the Government of Egypt.

The Khedive, in despair and disgust, offered Yr Majesty's Government to purchase his shares outright – he never would listen to such a proposition before.

Four million sterling! and almost immediately. There was only one firm that cd do it – Rothschilds. They behaved admirably; advancing the money at a low rate, and the entire interest of the Khedive is now yours, Madam...

The Government and Rothschilds agreed to keep it secret, but there is little doubt that it will be known tomorrow from Kairo.

Disraeli, in fact, was exaggerating the 'despair and disgust' of the Khedive. Other politicians, in particular Gladstone, not only thought that the interest rates were too high but that the queen did not have complete control of the Khedive's interests as his share certificates were pledged until 1894.

The British consulate in Egypt would be given the shares by hand once the money had been delivered by the Rothschilds to Ismail. This all went well, and at the end of November 1875 Disraeli's government was in receipt of 44 per cent of the Suez Canal Company, making them the biggest solitary shareholder. The Suez Canal by now was shown to reduce journey times not just among European nations but from New York to Bombay (by 31 per cent), Singapore (18 per cent) and Ras Tannūrah in Saudi Arabia (30 per cent). The British now held a global gem. The Rothschilds were praised by Disraeli in his letter to the Prince of Wales:

Our friends, the Rothschilds, distinguished themselves. They alone cd. have accomplished what we wanted, and they had only 4 and 20 hours to make up their minds, whether they wd, or could, incur an immediate liability of 4 millions. One of their difficulties was, that they cd. not appeal to their strongest ally, their own family in Paris, for Alphonse is *si francese* that he wd. have betrayed the whole scheme instantly.

The French press were furious. They declared a national scandal, foreseeing that this would jeopardise their concerns in North and Sub-Saharan Africa. De Lesseps had been eager to outbid Disraeli and was devastated to find that Britain, who had spent so much time denigrating his efforts, now owned such a large part of the canal. He could only hope that Britain and France would work together in the interests of business. The Germans, meanwhile, were very amused. But not all in Britain agreed with Disraeli's decision. *The Times* reported on the serious responsibility that such a large share would bring, warning that in doing so Britain was incurring a liability in the affairs of Egypt.

General Stanton arranged for the shares to be counted and found that there were 176,602 rather than the stipulated 177,642. The shares were contained in seven large zinc boxes and stored at the consulate, with two copies listing the shares and their numbers. They were shipped to Portsmouth on 16 December 1875 on the HMS *Malabar*, where they were transported and verified by the Bank of England.

The first payments, made on 1 December 1875, were to the London and Westminster Bank, the Bank of Alexandria, the Anglo-Austrian Bank and the Anglo-Egyptian Bank, to whom the Khedive was in debt. The Khedive asked Lord Derby for two Treasury representatives to help restructure the financial situation in Egypt, a subject that Stephen Cave was working on at the time.

Mr. Cave arrived in Cairo on 16 December 1875 with precise instructions from Parliament. However, the Khedive was alarmed at his presence, thinking that he represented other powers, specifically the Ottoman Porte. The British Foreign Office offered reassurances, and Mr Cave met with the Khedive on 25 December.

Egypt was truly falling apart. On 4 January 1876, General Stanton, the Consul-General in Egypt, reported on the offer by the Credit Foncier, Credit Mobelier and Bank of Paris of a loan of £12 million on a lease of the railway, telegraph and port of Alexandria for thirty years. There was an offer from Baron Hirsch of £2 million for six months at 5 per cent on the 15 per cent canal revenue, and also an invitation to form a company with a certain Sir George Elliot.

By 18 January 1876, Mr Cave had succeeded in presenting to the Khedive the general outline for a scheme for the funding of the debt of the country. This he intended to do by consolidating the debts and liabilities of the government and the Khedive, creating a government stock. It would terminate in fifty years, with a moderate rate of interest and a percentage going towards a sinking fund, which would be redeemed at the end of the period.

Mr Cave telegraphed Lord Derby to inform him:

> Referring to my Memorandum of 15th instant and my telegram of 21st instant, control proposed by the Viceroy of Egypt for the due payment of the annual charge for the consolidated stock is now restricted as follows.
>
> The Egyptian Government shall constitute a special administration independent of the general financial administration having a special superintendent and staff of employees. Of the two persons whom His Highness has requested Her Majesty's Government to send out to reorganise his finances would be the head of this new Control Department. Certain special revenues arising from land-tax, railways and harbours, sufficient to provide for payment of interest and sinking fund on the consolidated stock, to be paid into the chest of the new Department. Stipulations for assuring the full payment of the annuities, for the formation of a reserve fund and for the permanency of the English Superintendent at the head of the Department, are given. Viceroy of Egypt will include in the Decree an undertaking not to raise any other loan or contract, any other financial engagements which might cause a deficiency in the general revenue.

Mr Cave constantly communicated with British officials as to the state of play in Egypt. The government was a flurry of activity, with one MP reporting:

> On the 5th January [1876] Mr. Cave telegraphed that an opportunity of giving real support to Egypt ... existed, and

hinted that the form of support would be to render the Khedive assistance in getting the floating debt taken up by a responsible firm.

However, the British government remained cautious. George, based in Egypt, was in the thick of it:

On the 24th January Mr. Cave telegraphed that the consolidation scheme was being worked by Sir. Geo. Elliot and hinted that he might have to give provisional approval of it. The Govt. declined to hold themselves responsible for it, directed Mr. Cave to abstain from approval and asked for further details.

On the 25th Mr. Cave stated that he had already declined to approve Sir Geo. Elliot's scheme, and was prepared to suggest an alternative, but was recalled by a telegram from Lord Derby on the 26th.

The following day the annexed telegram arrived and was answered...by one from Lord Derby to the effect that the Govt. were prepared to name two gentlemen to take service in the Finance Dept.

About this date, His Highness received a proposal from Sir George Elliot and other English Capitalists based on a similar principle and stated that he intended to accept the proposal provided it met with Mr. Cave's approval, but Her Majesty's Government distinctly declined to approve of the Khedive's proposed plans, to express any opinion on them, or incur any responsibility in connection with them.

Meanwhile other offers had been made to His Highness, and negotiations were nearly concluded with a French Company for a loan of £14,000,000 at 9 per cent to pay off the floating debt, and for a further loan of £4,000,000 to cover a local debt.

On the 28th [January] Mr. Cave reported that the Khedive had nearly concluded a loan of £14,000,000 at 9 per cent from French contractors, that the Elliot scheme seemed at an end, and that H.M. Government were in no way compromised by what he had done.

Things did not stand still for long.

A later telegram the same day stated that the French loan had not yet been concluded, and that Elliot was again in the field, and a message from Lord Derby to Mr. Cave declined on the part of the Govt. to express any opinion on the Khedive's plan or incur any responsibility in connection with them.

H.M. Government appeared to consider that Mr. Cave was becoming too much mixed up in the Elliot scheme, and on the 28th January ordered his return.

However, Mr Cave's scheme to introduce two Englishmen to administer a new control department, independent of the general financial administration, was welcomed by the Khedive. It was reported that 'the French Government were desirous to lose no opportunity of increasing French influence at the expense of England'. George, for all his business acumen, could not gain his government's approval to finance his ventures. The British government were simply too wary about getting too involved in Egypt's affairs.

Considerable rivalry appears to have existed between the French and English capitalists but on the 6th February it appeared that the Elliot negotiations had failed through the impossibility of procuring an advance of funds, and the offer of the French from the exorbitant rate demanded ... On the 12th February it seems that a loan of £3,000,000 had been negotiated by French capitalists at 14 per cent to relieve the Khedive's immediate wants.

The French expressed regret at 'the antagonism between the French capitalists represented at Cairo by M. Pastre and the English capitalists represented by Sir G. Elliot'. George's business pursuits had made him well known in Egypt, with a newspaper reporting in 1876,

> Sir George Elliot Bart., M.P. for North Durham arrived at Ismailia on Tuesday, the 28th ult., and was received by the Governor and a guard of honour. A *cordon* of soldiers lined the route on both sides of the hotel, where Sir George made a brief sojourn, and on leaving again for the train, he was accompanied by a guard of honour, who presented arms both on the arrival of the boat and the departure of the train.

The wishes of the French government were to set up a joint commission, led by France, England and Italy:

> Lord Derby in reply on the 19th February pointed out that he did not see how the establishment of an International Commission could obviate the rivalry at Cairo, that H.M. Government did not desire to interfere in the internal affairs of Egypt, and were unable to join in any such measures for an international control over its finances.

The contract for the constitution of a national bank was signed that month by the Egyptian Minister of Finance and the managers of the Anglo-Egyptian Banking Company.

The debate in the House of Commons regarding the purchase of the Suez Canal shares commenced on 14 February 1876, with some cabinet members and MPs from the Conservative Party opposing their leader. Lord Derby was conflicted about the canal. He said that it 'had been originated by a Frenchman, executed by French engineers, and carried out principally by means of French capital', although four-fifths of the ships using it were British. He would have preferred the canal to be run by an international committee, but he did see the danger in it being taken over by a foreign power. Sir Stafford Northcote, the Chancellor of the Exchequer, and W. H. Smith, the Secretary to the Treasury, were also against the arrangement.

Disraeli predicted at the start of the opening of Parliament that 'there is to be a war to the knife when the Houses meet'. He was right. Disraeli faced huge opposition in the House of Commons. Some MPs were confused and thought that the Rothschilds had actually purchased the shares. Others thought the rate of interest charged by the Rothschilds was too high on a loan that was fundamentally secure. He counteracted this by saying that such a large sum of money could leave the banking firm out of pocket, especially if the Khedive had asked for it to be paid in gold. Other suggestions were that the money could have been gained from the Bank of England or voted by Parliament, but the counter-argument was the rapid timescale in which the deal needed to be made. Sir Stafford Northcote eventually came around, saying that the whole affair was completely out of the ordinary 'but so is the Suez Canal itself'.

Parliament eventually approved the purchase of the shares on 20 February 1876. The reception to the Suez Canal share purchase throughout the country and overseas was one of great joy. Having seen the cabinet minutes – which Disraeli had also conveyed to the queen – the press was enthusiastic at what they saw as a coup against the French. The city and industrialists were all behind the action. All over Europe leaders sent their compliments. The French newspapers were in favour of the British actions, especially in light of the Russian manoeuvres. Britain was now in a position to transact deals on the movement of their ships through the Suez, something that had caused intense conflict before. They still, however, only qualified for ten votes at shareholder meetings, and three seats on the board of directors.

Parliament was basically happy for the Suez Canal Company and the British government to be more closely aligned, enjoying the new balance between English and French concerns. Gladstone was said to have shown his true mettle, as he 'dwelt on the less important matters to do with the size of the commission and the dates on which the money should have been borrowed'. The world was happy – perhaps with the exclusion of Russia.

Russia at the time was intending to take hostile action against the Ottomans in Constantinople, with the view to protect the Christian communities in the Balkans, but the British sensed a threat of Russia gaining a foothold in the Mediterranean, which would hamper the British shipping route to India through the Suez Canal. Although there was a meeting in Bosnia of the main players – Britain, France, Italy, Germany, Austria-Hungary and Russia – no firm conclusion was reached.

W. H. Smith wrote to Baron Lionel, offering to pay £1.5 million on 10 March 1876, the same amount on the 20th and the remainder on the 31st, the total amounting to £4,075,996 13s 7d. Share prices had already risen from £19 with coupons at point of purchase, or £22 10s 4d with no coupons, to £34 12s 6d by January 1876.

The Italians expressed their willingness to join the French government's banking scheme and were happy to appoint a commissioner, but 'the Khedive expressed his anxiety for the nomination of an English Commissioner as essential to the success of the plan'. Mr Cave reported,

> A refusal by England would result in a breakdown of the Pastré arrangement and a financial crisis in Egypt. To this H.M. Government replied on 6th March that they could not send a Commissioner, but they would consider a workable plan for a Commission to receive revenues and apply them to payment of debt.

The Khedive, having borrowed most of his money from the French, was under considerable pressure to comply with their demands. The British did not want to get involved.

> The French group were very urgent, and threatened that, unless the Khedive that day assented to the conditions as to the Bank and Commissioners, they would not pay the £1,000,000 then due.

Twenty years earlier, when the canal had been seriously opposed by Palmerston, Disraeli had said, in his position as Chancellor of the Exchequer, that 'the operation of nature would in a short time defeat the ingenuity of man'. The journalist Lucien Wolf remarked that 'in nothing has British statesmanship blundered so badly as in its early attitude towards the canal'.

The French began to lose their cool. The British had been so opposed to the canal and thrown so many objects in de Lesseps' path, they declared; how dare they take over the majority of shares? The British government stood behind Disraeli's actions. However, Lord Chancellor Sir Stafford Northcote did communicate to Disraeli that he had acted in an underhand manner:

> Our policy, or our proceedings, with regard to the canal, had not been such as to gain us much credit for magnanimity. We opposed it in its origin, we refused to help de Lesseps in his difficulties: we have used it when it has succeeded ... and now we avail ourselves of our influence with Egypt to get a quiet slice of what promises to be a good thing ... I don't like it.

De Lesseps, nevertheless, was unperturbed. He liked the fact that it was the British and the French dominating the canal in line with his original vision:

> The powerful solidarity which had now been established between the French and British capital invested in the affair will ensure that this international Canal will be run on purely business-like and peaceful lines.

6

MP AND CAPITALIST

Three particularly significant aspects may be perceived in Elliot's career. First, by rising from the lower ranks of society to its peak through his own endeavours he represents quite strikingly the energy and confidence of the Victorian Age in the industrial field. Second, he played an especially important role in the development of the South Wales coalfield, exemplifying the change which occurred as groups of English capitalists bought the companies and coalmines from several venturous native Welshmen who had preceded them. Third, his political career as a self-made man was an excellent example of the new Tory Party which Disraeli wanted to create.

David Leslie Davies, 'Sir George Elliot',
Dictionary of Welsh Biography

This latter fact, George's meritocratic rise, earned him an audience with the queen, reported on 15 May 1875:

Her Majesty the Queen held a levee at Buckingham Palace on Wednesday, at which the following, among other presentations, were made the Queen:- Sir George Elliot, M.P., by the Right Hon. B Disraeli, M.P.; Mr George William Elliot, M.P., by the Right Hon. B. Disraeli, M.P.

He went on to have dinner with the Prince of Wales that evening:

> His Royal Highness the Prince of Wales honoured the Marquis of Londonderry by his company at dinner on Thursday evening at Holdernesse House, Park Lane. There were present to meet His Royal Highness the Russian Ambassador, Prince Dol, the Duke of Marlborough, the Marquis of Waterloo, the Marquis of Ormonde, the Marquis of Stafford, the Earl of Roden, Viscount Castlereagh, Lord Henry Vane Tempest, Lord Colville, Lord Carington, Lord Elcho, the Lord Chief Justice of England, Hon. J. Vernon, Hon. Francis Stonor, Sir H. Pelly, Sir George Elliot, Colonel Bateson, Colonel Stewart, Mr Delane, Dr Quin, Mr F. Johnstone, the Rev W. Rogers, and Major Russell, in waiting on the Prince of Wales.

After George had become an MP, his first meeting with Disraeli was reported in a contributed article to the *Scarborough Weekly Post* on Friday 7 June 1878. It reads:

> Never did leader of a great political party possess a more tenacious memory of individuals and events, or give proof of this memory in a more seasonable and cheery manner than the present Prime Minister of England. Just ten years ago Mr. Disraeli, then in opposition, was entertaining his followers at dinner. Among the company there suddenly appeared a newly elected member of the House of Commons. He was a stranger to many of those present; but he was no stranger to Mr. Disraeli, who stepped forward, warmly shook him by the hand, and said, 'Ah, Mr. Elliot, how many years ago is it since we smoked that cigar at Wynyard?' The gentleman thus greeted was the member for North Durham – then Mr., now Sir George Elliot. Wynyard was the name of the Durham estate of the Marquis of Londonderry, and 'that cigar' had, as a matter of fact, been smoked upwards of a quarter of a century since.

Mr. Elliot was at that time, and continued for several years subsequently, the head managing agent of the Londonderry mining property in the north of England, and it was at Lord Londonderry's house that he first made the acquaintance of the future Prime Minister of England, who, with that quick perception of character which pre-eminently distinguishes him, at once saw that Mr. Elliot had qualities which destined him to rise to a commanding position. If ever man succeeded in life, if ever man found himself in the front rank of a compact body of political followers, and could boast to the steady discharge of the plain duties of life, that man is Sir George Elliot. He has worked laboriously, un-intermittently from the first, he has waited his time, and success has come to him, as it comes sooner or later, in some shape or other, to all who know how to wait patiently, and work honestly.

Sir G. Elliot is not merely a very successful, he is a strikingly representative, man. Nor does he represent only that new power which modern progress has established amongst us – the wealth that is the result of science and enterprise combined: he represents in not less marked a degree the partial transition of political power in this country from a patrician caste to the people. This is a process which, while undeniably of a democratising tendency, has been effected with minimum of democratic result. While the foundations of popular power have been broadened and deepened in England, while we have passed Reform Bills, and finally arrived at household suffrage, we have had nothing of that levelling down which was at one time apprehended. On the contrary, the new rulers of England have been gradually educated in tone and feeling up to the standard of the old. It is really the aristocratic, and not the democratic, spirit which has been diffused at every step, and that this should have been the case is due in a large degree to the bestowal of titular dignities, whether peerages, or baronetcies, or knighthood, upon these men of the people who deserve well of the people's rulers.

Mr. Elliot was unanimously requested by the Conservatives of North Durham to allow himself to be put in nomination as their candidate on the eve of the general election in 1868. The representation of that division of the county had always hitherto been in the hands of the great families of the North exclusively – the Lambtons, the Liddells, the Vanes, the Williamsons, and the Chaytors. When the Conservatives came to Mr. Elliot they acted wisely and well, for they recognised the altered spirit of the age, and admitted into their calculations the new principle. A more genuinely Conservative step could not have been taken than this deliberate acknowledgement of the power deposited in the class of all that is best in which Mr. Elliot was the type. He had, by dint of great industry and great talents, already amassed great wealth. From being the manager of mining estates of others, he had come to be the proprietor of vast mining estates himself. He was the employer of labour on an immense scale in the collieries not only of Durham but of South Wales. He was a large shipowner. He was the partner in a great firm of wire-rope manufacturers, and it was in a very great degree owing to the skill and enterprise of Sir G. Elliot that the Atlantic cable was successfully laid. Many men have received a peerage for smaller public services than Mr. Elliot performed upon this occasion, and when it is remembered that, subsequently, Mr. Elliot spent a considerable fortune in party services to his own country, it is not surprising that we should hear of higher honours than the baronetcy – which might be described as due after the Atlantic cable was laid – as likely to fall to his lot...

In 1874 he found himself in a minority of nearly three hundred. But failure and defeat are not words to be found in the vocabulary of Sir George Elliot. He attributed his position at the poll to its true cause, gross, unblushing intimidation, if not absolute bribery...Sir George Elliot carried the day, and at a cost of nearly £80,000 – the election and the election inquiry expenses included – wrested the seat from the second and previously successful candidate. A baronetcy was a comparatively small reward for party loyalty of so practical and costly a kind.

The whole of these proceedings were full of anxiety and excitement, but they only served to bring out into strong relief one of the most remarkable traits in a remarkable character. Whatever Sir George Elliot may have felt, he never exhibited the slightest sign of decomposure, and he never did or said one bitter thing. It have been difficult for him to have had any more gratifying tribute to his charity and self control than the utterance which fell from the lips of a Nonconformist minister who, it may be said, voted against him. 'If,' he said, 'you want a model of a Christian man, there' – pointing to Mr. Elliot – 'he is.' It would be impossible to exaggerate the beneficent influences which the forbearance and conciliatory disposition of Sir George Elliot has exercised upon his own folk – the extent to which it has softened political asperities or improved the social relations between employer and employed. There are two other points about the man not less noticeable, the deliberate tone of his mind, which causes him well to weigh the consequences of action, hence necessarily to be slow in action, but never when action has once been taken to retrace his steps, or to regret that the opportunity for retracing them has passed; and, secondly, his loyalty to all those by whom he is surrounded – a loyalty in whose dictates he acts perhaps even more steadily than is quite consistent with strictly business principles. His friendship with George Hudson, 'the Railway King' through good and evil report – a friendship which ultimately secured for the colossus of the iron road a pension of £600 a year, and thus enabled him to end his days in comfort after long privation and suffering, and which has been acknowledged by his widow in terms of spontaneous and pathetic gratitude – is only one of the many instances of Sir George Elliot's fidelity to those who have once acquired the right to call him friend. As a member of Parliament Sir George has done much good and laborious work. He greatly helped Lord Aberdare in passing his Mines Regulations Bill. He has won the recognition for his knowledge and disinterestedness of the two direct representatives of labour in the House of Commons, Mr. Burt and Mr. Macdonald. He has made exceedingly useful suggestions in committee, when different industrial questions have been discussed.

As a shipowner he has been selected for special commendation by Mr. Plimsoll himself. Some of the good and useful works which he has done for the community at large have been mentioned. In addition to this he has no sooner won his lawsuit against the North-Eastern Railway, and acquired valuable proprietorial rights at Whitby, than he set to work to add a new watering-place to Yorkshire. His Egyptian enterprises, and the value of the new opening with which he has provided the trade of the kingdom of the Khedive by his harbour works at Alexandria, can only just be mentioned now. His character and career have been described by the Prime Minister himself in a sentence which is a masterpiece of terse and true eloquence. The two points on which Mr. Disraeli laid stress when he offered him a baronetcy were his services to the Conservative Party and his useful and honourable life; and if there is one addition which should be made to this group of virtues, it is his devotion to the Church of England, accompanied by a charity, as precious as it is rare, for those who are not in communion with that Church.

A banquet and testimonial was held in Sir George Elliot's honour, as reprinted by the *Newcastle Daily Journal*:

The great festive gathering, which was held in the city of Durham, on Wednesday, the 31st March, was as interesting in its purpose, and conception as it has been perfectly successful in its accomplishment. No man ever deserved better of a political party in a constituency, or was worthier of the honour bestowed on him than the recipient of the handsome testimonial which it was the object of the meeting to present. This is not the time nor the occasion when it is needful to enlarge upon the career of Sir George Elliot, or to show by what steps he has risen from the ranks to the position he now occupies...

When the General Election of 1868 was on the *tapis* it was understood that Mr Shafto, who had sat for the constituency from 1847, intended to retire. Hitherto the representation of the Division had been exclusively held by the county families; but the large additions made to the electorate by the Reform Act of 1867 suggested to both parties that the trading interest, of which the county had become the centre, ought to be represented. It would be difficult to have selected two more conspicuous representatives of that interest than the two new candidates, Sir, then Mr George Elliot, and Mr Isaac Lowthian Bell. Although they appeared upon opposite sides in politics, they were both gentlemen of whom the county which had produced them has reason to be proud. Mr Elliot put out his address several months before the General Election was expected to be held, and his appearance as a supporter of the Church of England and an opponent of the disestablishment of the Church of Ireland, and in favour of a measure of national education, and of the improvement of the condition of the industrial population. The great political question of the General Election was the disestablishment of the Irish Church; but other and much broader questions entered into the contest in North Durham. Mr Elliot's candidature was extremely popular, especially with the class he was proud to claim as his own, the miners, and he was also loyally supported by the old leaders of the Conservative party, who showed no disposition to stand aloof from motives of exclusiveness. After a very severe contest, it is well known that Mr Elliot was carried at the head of the poll by a very large majority.

His conduct in Parliament was a faithful fulfilment of his promises. In all Constitutional questions he acted staunchly with his party, but he was much more conspicuous for the part he took in the social measures of the Government; and especially in the Mines Regulation Bill. The author of that bill, who also had the charge of it in the House of Commons, Lord Aberdare, although a political opponent, has in the most cordial and honourable manner borne high testimony to the assistance received from Sir George Elliot in passing it. 'It is but bare justice to you,' said his Lordship, 'to say, that not only did I receive from you valuable advice with reference to the more technical parts of the Bill, but that you also displayed the warmest sympathy with the interests of the mining population with whom you were so closely connected, and to whose labours you were so largely indebted.' Higher testimony than this to the practical usefulness of a member in the position of Sir George Elliot, it would not be easy to conceive of, and up to the eve of the last General Election it was generally believed that he possessed as safe a seat as any member of the last House of Commons.

But although up to this period Sir George Elliot may be said to have been unusually fortunate in his political career, and to have worked honestly and faithfully to deserve the confidence and gratitude of his constituents, he was not destined to escape without a strong experience of the distance to which political passion will carry men on the road to injustice and persecution. The General Election was suddenly precipitated in his absence from the country, and at the same time a tremendous effort was made to carry both seats for North Durham by a *coup de main*. Sir H. Williamson was unceremoniously squeezed out of the field, and two gentlemen came forward in the Liberal interest, assisted, as was afterwards found, by an understood arrangement with the Durham Miners' Franchise Association, which secured them the operation of the whole machinery of the trades' unions in the division. Mr Elliot was not able to return home until within a few days of the election, and by that time his opponents had managed so well, that he was not permitted to show his face in the mining districts of the county without being exposed to violence. It is unnecessary at this time of the day to recall the slanderous misrepresentations which were so sedulously spread abroad, to the prejudice of Mr Elliot, and the exposure of which constituted his after triumph. It is only the simple fact, however, that Mr Elliot was denied by his opponents a hearing on his return from Egypt; and the election, which took place in the midst of a scene of violence and passion, resulted in the return of both his opponents.

Mr Elliot did not hesitate from the very first to declare that the election was an unfair one, and that he would do his best to procure for the constituency a more favourable opportunity of exercising their suffrages. A petition, duly lodged against the return of Messrs Bell and Palmer, was heard before Mr Baron Bramwell at Durham in the month of June last. For nearly a week the judge sat and received evidence of the rioting and violence which had attended the election at the various polling places throughout the division; and at length the evidence became so overpowering in its cumulative effect that the defence was abandoned, and the election declared void at Common Law. Up to this point Mr Elliot, who had in the interim between the February election and the hearing of the petition been created a baronet, a flattering letter from Mr Disraeli accompanying the bestowal of the dignity, might be said to have been fighting a battle principally in self-vindication. He had been deprived of his seat by a conspiracy of vituperation and violence; and he had exposed the scandal of the election and procured its nullification. At that point Sir George might have retired into private life, which we have every reason to believe would have best suited his personal inclinations. But the result, in all probability, would have been the return of both Liberal candidates; and nothing would have been gained for the Conservative party in North Durham.

There was no one in their interest ready to enter the field with any strong hope of achieving a victory against the powerful forces still arrayed on the other side. Under these circumstances, Sir George Elliot consented with the utmost public spirit to fight the battle over again. He had to do so in the face of an opposition ruffled and inflamed by the interruption of its success, and rendered the more angry by the exposure of the illegitimate weapons with which it had been fighting. The contest as we all remember, was as hot as a contest could be; but it was so far an improvement over that of February that the agency of the Miners' Union was kept in the back ground. Sir George patiently and resolutely combatted the slanders and misrepresentations of his opponents; and we have no hesitation in saying that before the end of the day it was rendered impossible for and honest and rational minded man to believe a word of them. The moral vindication brought a corresponding change in the attitude of the constituency for the polling restored Sir George Elliot to his seat, and indeed placed him within two votes of the head of the poll.

Sir George's troubles were not, however, past. It soon became noised abroad that there would be a retaliatory petition against him; and, although the election was known to have been conducted with scrupulous purity, the proceeding was all the more vexatious and none the less formidable. The petition was lodged within a few hours of the expiry of the twenty-one days allowed by law; and it was found to be subscribed by two obscure individuals from Sunderland. A counter-petition was promptly lodged against the return of Mr Palmer, and both petitions were ordered to be tried before Mr Justice Grove at Durham. The leading members of the Liberal party professed to have no responsibility connected with the petition against Sir George Elliot, but their legal agents were for the most part engaged in the prosecution of it. The charges bought against Sir George's agents proved to be of the most trumpery character, and after evidence had been led for two days and nothing of the slightest consequence elicited, the proceedings suddenly collapsed by the counsel for the petitioners withdrawing from the case. The petition against Mr Palmer was at the same time withdrawn.

[At the aforementioned banquet] – Speeches were given.

Sir George Elliot was determined that the seat that was won under such extraordinary circumstances of violence and intimidation to an extent hitherto unparalleled in this county – (hear, hear) – should not be wholly lost to the Conservative party – (applause) – and he spared no exertions to accomplish that object...he worked as no other man would have worked...He went about throughout the county, to all the polling places in the county, and to every little village to endeavour to disabuse the voters of the extraordinary ideas that had been put into their heads...It is our duty gentlemen to mark our sense of those energetic efforts on our behalf. (Applause) It is our duty gentlemen to do something to show the Conservatives not only in this county but throughout England that we appreciate the endeavours of a man like Sir George Elliot – (applause) – and that we are determined not to allow them to pass in silence.

Mr J. L. Wharton said, 'We wish to remain in friendship and harmony with those who have been our political enemies; but at the same time gentlemen, we do not wish to remain silent, or to remain unmindful of the eminent services which Sir George Elliot has given us in his constant endeavours to promote the Conservative cause in this country since 1868. (Applause) Gentlemen, Sir George Elliot has done, I believe, more in the way of promoting the Conservative cause than any man, as a candidate, throughout the length and breadth of this land. (Applause) I don't believe that any man who is now sitting in Parliament has met with such opposition – such organised, such powerful opposition – as, with all respect to our adversaries, I say Sir George Elliot has met with – (applause) – during his canvass...Well, gentlemen, I say that the whole of the Conservative party in the county of Durham owe its thanks to Sir George Elliot for his great and eminent services.'

Disraeli valued George's input and perspective on life, inviting him on 6 October 1875 to his country home of Hughenden. A newspaper reports:

Sir George Elliot, Bart., M.P., was one of Mr. Disraeli's guests on Wednesday, at the re-opening of the picturesque parish church at Hughenden, which has been restored at a cost of nearly £7,000. Among the other guests of the Premier were the Bishop of Oxford, Sir W.V. Harcourt, M.P., the Ven. Archdeacon Cust, Sir Philip Rose, Colonel Sturt, &c.

George had become an MP in 1868, representing the people of North Durham, although lost his seat as a consequence of the 1874 general election. He did, however, win it back later that same year, as told above. On 15 May 1874 he was honoured with a baronetcy in acknowledgement of his work for public services. In 1880, he again lost his seat in North Durham, regaining it in the by-election of 1881. This constituency was restructured in the Redistribution of Seats Act of 1885. George stood for Monmouth Boroughs and became the MP there in the 1886 general election, a seat which he occupied until 1892.

His granddaughter Florence Gordon recounts:

Sir George Elliot's political career was a phase altogether apart from his mechanical and commercial sphere. When the prospect of a general election presented itself in 1868, the Conservative of the old pre-reform constituency of North Durham looked around them for a candidate and succeeded in inducing Mr. Elliot to become the champion of their cause. He had never previously taken any active part in politics, but his leanings were strongly Conservative, and his personal qualifications were, with good reason, relied upon to secure him a strong following. It was one of Mr. Elliot's characteristics that he never forgot the rock from which he was hewn; and during his Parliamentary candidature he mixed freely and familiarly with the working classes of whom the constituency was so largely composed, and his unaffected geniality and frankness secured for him hosts of friends. The other candidates were Sir Hedworth Williamson and Mr. (now Sir) Lowthian Bell; and, although there was considerable disappointment, there was no great surprise when, on the declaration of the poll, Mr. Elliot was found to be far in front. The number of votes he received was 4,649, whilst his colleague, Sir Hedworth Williamson (Liberal), got 4,011, and Mr. Bell 3,822, the result being the gain of a Conservative seat. Although Mr. Elliot had thus added to his other duties the responsibility of a member of Parliament, he yet found time to devote to his ordinary business pursuits, and his engineering enterprises in Egypt were still further extended. Indeed, when Parliament was unexpectedly dissolved in February 1874, Mr. Elliot was at Cairo; and when he reached the scene of the electoral contest at North Durham, he found Mr. (now Sir) C.M. Palmer and Mr. Bell in the field and Sir Hedworth Williamson retired. He at once announced his candidature and was joined in it by Mr. R.L. Pemberton. The result of the polling was to give a majority to Mr. Bell, who had 4,364, and to Mr. Palmer, who had 4,327 votes; while Mr. Elliot had 4,011 – precisely the same number as was given to Sir H. Williamson when he stood second to Mr. Elliot in 1868 – and his colleague, Mr. Pemberton, had 3,501. Mr. Elliot, therefore, found himself defeated. It was on this occasion that Lord Aberdare in his letter to Mr. Elliot stated his opinion that during the progress of the Mines' Regulation Act through Parliament, and during the preparation of that measure he frequently asked and always received from Mr. Elliot very valuable assistance.

To the Conservatives the defeat was the more disappointing from the fact that their party had now practically acceded to power in the state; but the gallant exertions of Mr. Elliot did not go unrewarded. On April 20, 1874, he received a letter from Mr. Disraeli, (afterwards Lord Beaconsfield) intimating to him that the Queen had been pleased to confer a baronetcy on him 'in recognition of his great services to the Conservative cause, and of the enterprise and energy of his useful and honourable life,' and the dignity received consummation by its publication in the '*Gazette*' of May 1 following. Meanwhile, there had been murmurs of discontent as to the manner in which the late contest in North Durham, had on the Liberal side been conducted. This feeling found vent in the filing of a petition, and the trial resulted in the unseating of both members. On the sixth of the seven days over which the trial extended, the Judge plainly indicated that unless rebutting evidence of a strong character was forthcoming the effect of his judgement would be the unseating of the two members. Acts of intimidation, for which the Liberal agents were responsible, were clearly proved, and the election was declared void. A new writ was issued in June 1874, and with the sitting members, who had been personally exonerated from blame, Sir George Elliot once more entered the lists. The election on June 19, although characterised by extraordinary exertion on both sides, was conducted with perfect order. The result of the polling was to restore the Conservative to the position from which he had been ousted. Mr. Palmer, and the other candidate, Mr. Bell, polled at 4,104.

Florence continues,

In celebration of his triumph, and in gratitude for past services to Conservatism generally, the hon. baronet was, on March 31, 1875, entertained at a banquet in Durham, under the presidency of the Marquis of Londonderry, and was at the same time made a recipient of a magnificent service of plate, valued at £3,000. On the dissolution of 1880, it seemed for a time as if Sir George was to be allowed to regain his seat without challenge, but at a somewhat late stage of the electioneering campaign, a second Liberal candidate was announced in the person of Lieut. Col. Joicey. The contest, though short was keen, and ended in a victory for the Liberals, with the following figures: – Joicey (Liberal), 6,233; Palmer (Liberal), 5,901; Elliot (Conservative), 5,092. In the following year, Col. Joicey died, and there was a bye-election, at which Sir George received 5,548 votes, and his opponent, Mr. Laing, 4,896. Sir George represented the division until 1885. After the redistribution, he contested the South-East Division against Sir Henry Havelock-Allen, the result being that Sir Henry defeated Sir George by 5,603 votes to 4,854. This practically ended Sir George Elliot's intimate connection with the North-Eastern Counties, and he transferred the closer relationship to Wales, where his interests were very extensive. In 1886, he was elected Conservative member for Monmouthshire Burghs, receiving 3,033 votes, while the Gladstonian candidate, Mr. E.H. Garbutt, got only 2,568. In 1892, however, at the general election Sir George was defeated by Mr. Albert Spicer, Gladstonian. Sir George Elliot's career in Parliament, though not brilliant was consistent and beneficial. On all labour questions he spoke with authority, and in the amelioration of the condition of the mining population he was deeply and actively interested. He was no orator but spoke only when he had a knowledge of the subject under discussion; and his sentences, halting perhaps, were yet characterised by a shrewdness and a common sense that gave them weight. Those who have heard him address public meetings cherish a remembrance of a smiling face, lit up with a genial good nature, and a manner full of a natural charm, that won for him the esteem and attachment of all men.

George was extremely absorbed in politics, but he was also an active businessman, and sometimes these interests clashed, as Florence explains:

> We have referred already to Sir George Elliot's enterprises in Egypt, these being the Harbour of Alexandria with which are identified the names of Sir George Elliot and Messrs. Greenfield. The advantage to shipping was incalculable, the new harbour comprising 1,400 acres, hence his sudden return to Egypt in 1874. Then and afterwards he continued to take a close interest in the physical and financial improvement of the country. In 1875 and 1876 he was again in Egypt, and on January 27, in the latter year, he was received in special audience by the Khedive, the Ministers of Finance and of Foreign Affairs being present at the interview. The object of his mission was to arrange on account of Messrs. Greenfield and Co., certain matters then pending in respect of harbour works, and a suggestion was made to him as to the procuring of a loan of £2,000,000, on the security of the harbour revenues. Sir George expressed to the Finance Minister his conviction that this partial and temporary mode of dealing with the finances of Government would be only to prolong a system which had already involved the country in great pecuniary embarrassments, and that some more comprehensive and effectual means should be adopted for readjusting the finances and restoring the credit of the Government. Sir George Elliot, by the invitation of the Government, prepared a project, the principle of which was the consolidation of the debts and liabilities of the Government and of the Khedive, by the creation of Government stock, into which these debts and liabilities should be converted; and in accordance with at least a portion of the scheme, the Khedive decreed the consolidation of his debt; £91,000,000 at 7 per cent. and a sinking fund.

George communicated with the British government on this point, with special reflection on Alexandria Harbour, with which he was so involved. British government correspondence on this matter survives:

> Upon this point, Sir George Elliot, the contractor, raised some objections on the ground that it was expedient to keep the control of the port in English hands; but his objections were met by the fact that his contract remains undisturbed and that at its expiry it is left open to the Viceroy to make a more favourable arrangement, if he can do so, for paying off the £2,000,000 and releasing the port dues from this mortgage.

In telegrams dated 5–10 January 1876, George communicates with his son-in-law, the industrialist and journalist Joseph Charles Parkinson of 13 Victoria Street, about this matter and others:

> Propositional lease for 30 years all the Egyptian Railways, 1,300 miles long, with whole of the Rolling stock, stations, and every other adjunct net earnings, 1874, were [£]990,000; second lease of the Mahmondik Canal for 30 years to prevent competition with Railways, net earnings 1874, were [£]25,000; third tobacco duty or tax amounting to [£]250,000 and increases very much, the customs due on spice liquor [£]8,000 capable of tenfold increase after 2 years. These all amount to 1,273,000 pounds to which it is proposed in the future to add receipts for dues from Government hitherto exempt and amounts to about 150,000 pounds sterling per annum. There is also the Port and Harbour of Alexandria, surrounded by 70 acres of reclaimed land, navigation complete, net revenue of which is estimated at [£]200,000 by immediately realizable 150,400 pounds sterling per year guaranteed or the price to be modified should on investigation any errors be found in the statements. The Loan of [18]73 is entitled to [£]750,000 of this say from Railways, leaving a balance over [£]650,000 to be augmented immediately by charge on Government traffic of [£]150,000 in addition to the increasing production of the country estimated at

6% per year, and the economy to be effected in working Railways, Port etc. What is all this worth, the solution is urgent, as the floating debt [£]12,000,000 to be met during year [18]76, the capital required from the Company ought not to be very large, debenture at 5% or 5 ½ % with sinking fund, could be issued and ought to cover ¾th and ordinary share would receive the profit; ascertain if any terms could be offered. Khedive has promised to one and Consul General [General Stanton] a delay of 10 days for any offer or for any politic bonafide negotiation. He prefers the English. French in competition. Lesseps here.'

He writes further to Parkinson on 5 January 1876:

Foreign office yesterday officially enquire about this affair and has been informed that French are treating; to enable me to communicate with England a delay of about 10 days has been offered to me by the Khedive. Unless Government, Rothschild or some strong combination embrace this opportunity and secure this hypothecation it will pass into other hands, and English political and commercial interest will suffer.

Disraeli's secretary, Montagu Corry, responded with a tentative offer for the Port of Alexandria, which the Egyptian government refused. They wanted in the region of £2,250,000 – an impossible sum.

Rumours were flying around as to the situation, with the French and other authorities also bidding for these assets. George telegrammed London on 5 January 1876, responding to gossip regarding Mr Cave, the British representative in Cairo:

The rumour as to Mr. Cave leaving or being dissatisfied or any misunderstanding with Khedive or Finance Minister. I am after communication directly with both as well as my own personal knowledge of what is transpiring here. Authority to say that there is not in the least foundation in the rumour and … Cave has informed me that his mission so far is perfectly satisfactory.

International financiers were like bees to honey. They swarmed in on Egypt with offers for tobacco duty and railway revenues subject to the 1873 loan offering £1 million in return for these with a sinking fund of 7–8 per cent with equal division on the profits. George telegrammed London again on 5 January 1876 asking for the government's opinion. Events were moving fast.

His telegrams were addressed by Mr Parkinson, who says on 6 January 1876:

> Have consulted Corry, Mills and Oppenheim. Securities offered seem absolutely inadequate for finding any amount approaching [£]12,000,000. In fact, in present state of Egyptian credit and public feeling which has been greatly shaken, any operation unless done under guidance of Cave extremely difficult, if not impossible. We are conferring and will telegraph tomorrow.

Egypt was being torn apart by creditors and it seemed that everyone wanted their piece of the pie. George's telegrams were shown to Disraeli, Lord Derby and Sir Stafford Northcote. They were uneasy about taking any further action until Rothschild had heard directly from Mr Cave, as similar proposals had been made by Egypt to Paris in the last fortnight. They felt that the Khedive was playing one country against another to escape proper control, and that French capitalists would not help unless they loaned Egypt money at exorbitant interest. The British government expressed doubts as to whether the 1873 loan could be secured by the railways, which they felt was limited to £730,000. Their feeling of general disappointment and distrust in Egypt was not appeased by a published contradiction of the Egyptian government's quarrel with Mr Cave; in the British government's words, 'only Cave can restore confidence'.

The whole thing was a mess. Parkinson reported back to George on 7 January that he had met members of the Ottoman

Bank. He confirmed his feelings of uneasiness in dealings with Egypt, communicating to George that 'feeling is greatly changed since you were in city. Arrangements contemplated then would be impossible now. *Times* this morning had leader damaging Egyptian Credit.'

The exact working of the 1873 loan was called into question and George received an original copy to examine. His colleague Lloyd was unsure of whether all the earnings from the railway were pledged instead of £750,000 only, which the Egyptian government claimed. Lloyd looked closely at the document and he was clear on the fact that bond holders' rights were not limited to £750,000 but were entitled to the majority of the proceeds, except a small portion of the Lower Egyptian Railways. Therefore, the Egyptian government could not guarantee any advance from Britain as they did not have the income to assure it.

George goes on to say that he thinks, given the circumstances, that the Egyptian government should consolidate their loans, estimated at nearly £100 million and accruing interest daily, with the unequivocal condition that these debts should be in the hands of England or another European state appointed by the bond holders. He suggests a sinking fund of forty to fifty years, at a rate of interest determined by the present market price. He concludes, in his letter to Mr Parkinson, which was shared with Disraeli and Lord Derby:

In the extremity of present crisis and prospects, I have reason to believe that this will be agreed to; give these my suggestions early and thorough consideration; so large a proposition probably ought to be done in combination by England and France, I should prefer English. My impression is that in present state of affairs which cannot last, but which may be tidied over that the larger scheme now proposed is the alternative to prevent Bankruptcy. I am in personal communication with every person of note both English and Egyptian, my assurance from all points is a preference to be in English hands.

Hardly surprising considering the rate of interest the French were charging on their loans.

George wanted to secure his investments in Alexandria Harbour, which his company, Greenfield and Co., was developing. Harbour dues to Greenfield and Co., at the Liverpool rate of 6 per cent, created ample security on the £2 million already expended there. George himself had invested £750,000. It was a bit of a sticking point for George, as Greenfield and Co. were only entitled to possession of the harbour until 1879. He describes Alexandria Harbour as a 'stronghold, with 1,400 acres of enclosed water, the greater part of which is deep enough to receive the largest Ironclads, and within 12 hours of [the] Port Said entrance of the Suez Canal'.

George wrote extensively from Cairo to the British government in January 1876. Mr Cave, working with the Khedive agreed:

> If the figures which have been placed before me are accurate, the security offered by Khedive and Finance Minister – the Khedive's 15% of the net profits of the Suez Canal, the Revenues of the Harbour of Alexandria ... and the Tobacco Duty amply sufficient to cover a loan of 2 million sterling.

George had worked out that this sum would pay off the liability of the Egyptian debt in February and March, and it would give them time to work out the larger scheme. He had been told that the rate of interest would be 7 per cent and was happy that this would work, but if it was absolutely necessary, he said that it could rise up to 1 per cent more. He was working on a project that would enhance the railways and harbour and stated that 'if the money is found much benefit will result to those interested besides conferring great Benefits on Egypt and keeping England paramount here'.

He goes on to say,

> Cave, General Stanton and Lloyd are equally anxious with me in the present crisis as we know the French are clutching

at this great chance and making offers which if we fail will be accepted. Exercise your discretion in showing this and let me hear from you affirmative. See Corry, and with him Rothschild and others, money required in ten days at latest – show this to Montagu Corry.

By 29 January, word in Downing Street was that the French scheme had failed. George pressed ahead, with pressure from the Khedive to get a speedy answer. He did not give up, writing to Disraeli on 12 April 1876 from his home in Park Lane:

Pondering over the much discussed subject of Egyptian finances, and appreciating fully the many difficulties that stand in the way of the direct intervention of the English Government, the following had occurred to me as a possible mode of dealing with the question.

He writes that he had had a long conversation with Baron Rothschild that morning regarding this, and that it 'transformed ... our general discussion on Egyptian finance'. The *Standard* reports on 12 April 1876:

Sir G. Elliot stated that for many years he had periodically examined the Suez Canal, and had seen it every year since it was opened. Having been there in January last and examined its condition, he was of opinion that there would be no necessity to spend thirty millions of francs upon it, and that we had got a really good and substantial property in the purchase of the Suez Canal shares for four millions (hear, hear). The income of the canal was increasing at such a rapid rate that he estimated that at the end of the period of nineteen years our four millions would be worth eight millions to this country. Indeed, he had no doubt of it (hear, hear). The canal could be kept in perfect order for less than 1000*l.* per mile or 90,000*l.* a year. He was acquainted with the engineers who would undertake the maintenance of the canal at that amount.

It should not be forgotten that the expense was not increased by the amount of work done, for a canal differed from almost any other work, as there was no wear and tear upon it from extra traffic. He saw no reason why the canal should be doubled, and in fact 45 miles of it was already practically double. A very slight expenditure would enable the canal to do double the work. The amount of the expenditure last year was nearly 1,250,000*l.* It appeared to him almost an act of grace that this country should have three representatives upon the direction of the canal, and if the English people wanted more interest let them acquire more shares, for shares were to be bought in the market every day. He should be glad if he could impress on the government the importance of the idea of carrying out their purchases still further (hear, hear) while there were shares still in the market (hear, hear). He would strongly recommend that the government should acquire an interest which would enable them to assume a position of more importance. He trusted that this suggestion would be deemed worthy of consideration by the government.

George made numerous financial suggestions to the government, but the politicians in question were very wary of what they are getting into given the unsteady state of Egyptian affairs and their political relationships with both Russia and Turkey. There were two conflicting views in British politics on where their national interest lay; was its support for the Turks, who had a record of ruthless brutality towards anyone they didn't like, or for the Russians, a major competing world power, who wanted to occupy Constantinople to give access to the Mediterranean, thereby threatening British shipping?

Tensions were high as ever. Montagu Corry, Benjamin Disraeli's private secretary, 'could give no answer beyond that Mr. Disraeli had received Sir George's communication'. He says,

I then remarked for myself, on the wildness of the notion that the Prime Minister would ask a firm of financiers to float a scheme in the market on a foreign state – and that, in my case (as Mr. Disraeli had observed) Sir George Elliot's was no *proposal*, but only a *plan*.

The *Observer*, in April 1876, reports on Mr Cave's report and Sir George Elliot's scheme:

Viewed in its true character, as the statement of an agent sent out to advise his principals as to the possibility of extricating an embarrassed estate from its financial difficulties, the document is one of singular ability and value. No doubt a question may be fairly raised as to the adequacy and accuracy of the information on which Mr Cave based his conclusions; but a study of the report can hardly fail to convince the candid reader that it was a result of very careful and intelligent investigation. Moreover, the correctness of its conclusions is confirmed by the fact that – as was shown by Sir Geo. Elliot's report – an independent investigation, conducted at the same time by third parties from an independent point of view arrived at a substantially identical result. And granted the correctness of Mr Cave's premises, it seems to us that the report is far more satisfactory in as far as Egypt is concerned that could reasonably have been expected. That the Khedive was under the pressure of extreme financial embarrassments was known before Mr Cave started on his journey, and was indeed the avowed cause of the mission itself. It can hardly be said that these embarrassments have proved on inquiry to be more serious than was anticipated; while on the other hand nothing can be more decisive than Mr Cave's opinion as to the 'sufficiency of the resources of Egypt, if properly managed, for meeting her liabilities.' Mr Cave's report, in fact, may be fairly summarised in the following way:- Egypt, viewed as an estate, is rich enough to pay 20s. in the pound; but the immediate liabilities of the owner of the estate are so onerous that he cannot possibly meet them out of his current revenue, and therefore bankruptcy must ensue unless he can raise money upon mortgage to pay off his outstanding debts. At the same time, no mortgage can be safely affected unless adequate security can be given that the net revenue of the estate shall be applied in the first instance to the payment of the interest, and that no fresh liabilities shall be incurred till the debt is paid off. It is on the basis of these conclusions – the substance of which we communicated to our readers many weeks ago – that Mr Cave grounded his proposal. His scheme died stillborn, owing to the drawing back of our Government from the logical consequences of their own action. But it is worthwhile to point out the essential characteristic of this scheme, namely, the establishment of a control department, whose authority would be virtually, if not avowedly, independent of the Khedive's favour. This idea, though expressed in a more guarded form, lies at the bottom of the Elliot project, and we believe ourselves that it must be the fundamental principle of any effective regeneration of Egyptian finance. How far the scheme now brought forward in France fulfils this question is a subject we have discussed elsewhere. For our present purpose, it is enough to say that it would have been fulfilled by Mr Cave's suggestion of placing at the head of the control department a financial agent sent out by the British Government, and who, though he would have been placed in his post by the Khedive, could not have been removed from it without the sanction of the Power by whom he was nominated.

George carried on with his scheme, writing further to Disraeli on 14 November 1876. He talked about the Khedive's decree of 18 October that year, to which George had added various clauses. George was still working to keep the harbour under British control and wanted the Prime Minister to pay the Khedive £2 million to secure it, and not just because Alexandria Harbour was key to English national strength. George was able to insert various clauses into the Khedive's decree, namely clauses 3, 5, 24, 27 and 30, which dealt with Alexandria Harbour, ensuring what he wanted – that money due would go to English contractors and to ensure that the harbour and port of Alexandria would come into the hands of either an English company or the English government after his contract ran out.

However, he did admit, 'events seem to develop themselves rapidly, it may be too much so'. Downing Street dug in their heels. George got what he wanted, but there was a barrage of communication between governmental officials who were exasperated with him. Indeed, George had written to Disraeli on 19 August 1876 offering his 'humble congratulations' on his peerage, as Disraeli had just been made 1st Earl of Beaconsfield, and was therefore entitled to sit in the House of Lords. George goes on to say in his letter:

> At the same time, I am bound to add that the House of Commons has now ceased to command my enthusiasm. The only figure of all others that I cared most for is now removed from it. Henceforth my anxiety to be counted amongst its members will be increasingly abated.

Whilst showing extreme loyalty, these were fairly strong words, and George meant them. The tiresome tittle-tattle in Westminster about his Egyptian plans was testing his patience and he wanted to the government to be clear about his role and support his endeavours, especially in view of the fact that he was working closely with the Khedive. The British government generally felt that George was looking after his own interests rather than those of the nation, but perhaps to George they were one and the same. The government would not give clarity on whether George would secure the harbour after 1879 and was unsure how to credit him, with one official remarking on 18 December 1876:

> It seems to me clear enough that Sir George Elliot who is the contractor and Mortgagee of the Alexandria Harbour Dues has been looking, very naturally, after his own interests

and that his patriotism was aroused by the danger which he believed threatened his purse ... I do not see anything for the Government to be nationally grateful for.

But George did not give up. He wrote to Disraeli on 9 November 1877, nearly a year later, telling him,

Mr Vivian, the Consul Francais of Egypt is now in London, but is obliged to leave on Tuesday morning for Egypt ... I have spent some time with him this morning and have no doubts your Lordship would be interested in seeing him.

He wrote several more times to Disraeli, pointing him in the direction of people well versed in Egyptian matters who would be able to advise the Earl of Beaconsfield further. He also wrote to the government on the question of the two commissioners who were to be appointed by the English government to advise on the Khedive's debt as he was concerned for the security of his company, Greenfield and Co., and wanted to make sure that they did not 'interfere with validly special securities existing in virtue of such conventions and their execution'.

Meanwhile, he was enjoying himself in Cairo, hosting a banquet that was reported via submarine telegraph in 1878:

Mr. Stanley, the African traveller, has arrived here, and is to be entertained this evening at a grand banquet given in his honour by Sir George Elliot, Bart., M.P. The principal English and American visitors and residents and a considerable number of Egyptian Pachas have accepted Sir George Elliot's invitations, and this Anglo-American New Year's Day entertainment is expected to be one of the most brilliant and noteworthy incidents of the Cairo season.

The Ottoman Empire requested help from Great Britain in 1878 to prevent Russian aggression, in pursuance of the Anglo-Turkish

Convention. Cyprus was ceded to Great Britain and Sir George Elliot accompanied the First Lord of the Admiralty and the Secretary of State for War on board HMS *Himalaya* for their official inspection of the island. Although he acted in a purely private capacity, he was of much service in testing the practical worth of this new possession.

George also received an honour: the Grand Cross of the military order of Our Lady of Villa Viciosa, given by the King of Portugal.

Despite Disraeli's appointment to the House of Lords, George's relationship with the world of politics was far from over. He had acquired a great deal of land in Whitby, North Yorkshire, and on 23 July 1879 he invited Montagu Corry to stand for Parliament in that seat. Corry turned down the offer, even though George was willing to pay all the expenses. George made his feelings clear: 'I think the seat which I have occupied now for 12 sessions has ceased to be a subject which I ought to aspire longer to occupy but might make way for a younger man.' He goes on to offer his assistance to Lord Londonderry's son, Lord Lambton, who had lived in the North Durham constituency since 1800:

> I may mention that should I decide to remain in Parliament and should my suggestion about North Durham be adopted, I believe that I could be elected at Whitby without a contest and thus gain a seat.

Lord Londonderry also declines his offer:

> I feel very much obliged, for your kind expression towards myself and my son, Lord Henry. But I think it right to state, that I have no intention of bringing him forward for the County of Durham, and indeed I know that he has no wish, being fond of his profession, to enter Parliament, and having

I hope received the seat for my eldest son, I am unwilling to enter further with my Political struggle.

George was worried about politics in the north. He wrote to Montagu Corry on 27 November 1879, concerned that Radicals might fill the six seats available in North Durham after he had stepped down. He discussed the option of making Durham University a separate seat, which would probably vote Conservative, and advised on men who could possibly take it. The following day, George wrote again to Corry with the news that Lord Durham, having been dangerously ill, had died the previous evening. George felt that this would weaken the cause of the Radicals in the county.

On 29 November 1879, George received an invitation to go down to Hughenden, the country home of the Benjamin Disraeli, to talk further on this matter. He replied to the Prime Minister, saying, 'I have the honour to express my gratitude to your Lordship for such a distinguished [mark] of friendship. I propose to leave by the 4.45 [train] on Monday 1 Dec arriving at six.'

Events in Egypt were not faring well. Ismail could only last for so long on the £4 million he had received for his Suez Canal shares. Within a year he was again short of money, so much so that an Anglo-English commission, known as the Dual Control, took over his country's finances. Both countries had eventually decided to cooperate. Ismail was deeply humiliated. His plan of freeing Egypt from the Ottoman Empire and his ambition to modernise the country to attain European standards had failed completely. The Dual Control that had taken over the treasury was backed by many of his ministers, who realised that Ismail's ability to rule had totally floundered. Eventually, in April 1879, the incoming Turkish Sultan got rid of him. Ismail went into exile in Italy, while his son Tewfik took over the reins. The Dual Control sold the Khedive's 15 per cent of the dividends of the canal, and in doing so removed all Egyptian interest in it.

It was the Egyptian people who were most hurt. Thousands of fellahin had been used to build the canal, and Egypt now had no stake in it whatsoever. They would not even receive the profits from their shares. Tewfik was not a strong leader and allowed both British and French control over his government. This incensed the heavily taxed Egyptian people further still.

Within a few years, in 1882, the country reached breaking point. Colonel Ahmed Urabi, the government's head of war, led an uprising against Tewfik and the foreign powers that directed him. The Egyptians supported him in his revolt for 'Egypt for the Egyptians'. They were fed up with the benefits enjoyed by foreigners: tax breaks, their own justice system, and huge amounts of money flowing out of the country. A political crisis ensued, then came the military coup. Tewfik requested help from both the British and French governments. Britain responded, and in the summer of that year British boats bombed Alexandria, put to shore an army and crushed Urabi's troops at Tel al-Kabir. The French did not join them.

With France absent, the British were poised to take over the entire country, the canal included. De Lesseps was furious as he had always intended for the canal to be neutral, used by ships of all nations, but in August 1882 British forces took over the headquarters of the Canal Company in Port Said and quelled rebellious elements there. They later stormed Cairo, commencing seventy-five years of British rule in Egypt.

The French still nominally ran the Canal Company, but the British had complete jurisdiction. Most of the ships passing though were British. It was dredged regularly to prevent the build-up of sand and enlarged to allow several ships to pass at any given time. *The Times* was right. Britain did have 'an abiding stake in the security and welfare of Egypt' when it bought out the Khedive's shares.

The British used the Suez Canal as a pivotal point in their empire, allowing for development of trade and territory. Britain was extremely sensitive about protecting the canal, and as the twentieth century dawned they would expand their dominion into Afghanistan, the east coast of Africa, Iran and parts of the Middle East. They feared that rivalling nations such as Russia, France and Germany might take over these lands, putting the canal at risk. The canal was serving a vital role in connecting the British Empire, joining England with its distant outposts in Australia, Singapore, India and Hong Kong. They would hold the controlling interest in Egypt until after the start of the First World War.

Around this time, an edition of the *Illustrated London News* dated 21 July 1883 had a long article on the Suez Canal. It contains a map showing 'Sir G. Elliot's Freshwater Canal', running from the city of Alexandria to Ismailia. A further canal, labelled as 'Elliot's Parallel Canal', runs alongside the Suez Canal, from Port Said to the Suez. Clearly George had an abiding interest in the area.

Back in 1880, Ferdinand de Lesseps had gone on to inaugurate his second canal company with his son Charles, this time working in Panama in Central America. He operated on the same model as before, selling shares to the people, and planned for the canal to be open by 1892, but there were many hurdles to overcome. Thousands met their death in the heat of the disease-ridden jungles. As first believed with the Suez, but this time correctly, there was a difference in sea levels between the Atlantic and Pacific, meaning more digging was needed, as well as locks. By 1889, with almost 1 billion francs consumed, the Panama Canal Company went into liquidation. The directors were arrested and thrown into jail. Both Ferdinand and Charles de Lesseps were given a five-year prison term. Ferdinand de Lesseps was spared due to his advanced years – he was eighty-seven – and an illness that kept him bedridden in his country manor. He would die a few years later, on 7 December 1894.

The safety of the Suez Canal was enshrined in an international treaty on 29 October 1888 when officials from Britain, France,

Italy, Spain, Russia, Turkey, Germany and Austria-Hungary signed a document declaring:

> The Suez Maritime Canal shall always be free and open in time of war as in time of peace, to every vessel of commerce or of war, without distinction of flag. Consequently, the High Contracting Parties agreed not in any way to interfere with the free use of the canal, in time of war as in time of peace. The Canal shall never be subjected to the exercise of the right of blockade.

Three years after the treaty was signed, and before he passed away, de Lesseps' statue was erected at Port Said. One arm reached out to the canal, and the legend beneath his likeness bore that motto he had always used: '*Aperire terram gentibus*', 'To open the earth to all mankind'.

The Panama Canal eventually achieved success when an American company took it on, at huge cost. The Suez, meanwhile, continued to be strategically important. Over the course of the First World War, the British government placed over 100,000 soldiers in the area and instigated an Arab revolt, which was to have the dual purpose of safeguarding the territory and compromising Turkey. The number of troops in the canal zone remained high after the First World War, when Britain ruled Palestine, Jordan and Iraq.

Egypt regained its independence in 1922, with King Fuad I rising to the throne, but Britain still kept soldiers in the Suez region. As time wore on, Britain allowed the Egyptians more control of the canal and started to pay a yearly lease to the country for its use. However, for the duration of the Second World War Britain once again controlled Egypt. The canal was a high priority for the warring parties, using planes to bomb canal shipments bound for India and Asia. King Fuad I was deposed on 23 July 1952, and some years later Colonel Gamal Abdel Nasser took control.

The concession for the Suez Canal was due to complete its ninety-nine-year license on 17 November 1968. The Canal

Company brought in a huge amount of money for the Egyptian government, while the company still acted as a self-governing state. The canal was used extensively for the growing transport of oil from the Middle East. In 1950, out of a total freight of 122 million tons, some 75 million tons was oil.

In 1956, Nasser's application for credit to create a huge dam at Aswan, on the Nile, was rebuked by the World Bank and the Eisenhower administration in the United States. In July of that year, he broadcast a message that was heard all over the Arab nations. He talked about the conflict with European powers and the magnanimous triumph of the 1952 regime change. His speech fired the enthusiasm of the Egyptian army, whose troops flooded into the canal region. To the shock of the incumbent forces, the Egyptians took control.

By October 1956, Britain, France and Israel were undertaking military action against Egypt. Huge areas of Port Said were bombed, and British troops managed to regain substantial regions of the canal. The Egyptians responded by exploding the 35-foot statue of de Lesseps. The deciding factor in what was termed the Suez Crisis came when the United States condemned the actions of their erstwhile allies. They would not condone the military operations taken by Britain, France and Israel, and without this support the three countries pulled out. Egypt won its freedom. For Britain in particular this was a massive subjugation, heralding the end of the empire. Anthony Eden, the shamed prime minster, handed in his notice.

As global politics changed, and the Cold War became more predominant, the canal's immense status faltered. However, its revenues provided a sorely required renumeration for the Egyptian people over the course of the following ten years. In 1967, Egypt saw further battles with Israel. Port Said, Ismailia and the Suez were all but destroyed. As a result, the canal was impassable until 1975, when Egypt and Israel signed the Sinai Interim Agreement peace treaty on 4 September in Geneva. The canal was again open for business, and dredged to prepare for the new, bigger oil tankers, but its fortunes were to take a turn for the worse. By the early 1980s, oil was more frequently transported through pipelines, and the canal was not deep enough for the latest tankers. Businesses started to use other routes, and the Egyptians responded by lowering their prices. By 2000, the canal was bringing in US$2 billion annually – still a significant amount.

7

FAMILY AND FRIENDS

The *Illustrated London News* recounted on 15 May 1875:

Amongst the remarkable 'self-made' men whom the counties of Durham and Northumberland have produced is 'Geordie Elliot' as his countrymen familiarly call him. He has risen from the ranks, as George Stephenson did, by the patient and preserving exercise of rare talents and indomitable will. Sir George Elliot, M.P. for North Durham, was born at Gateshead, in 1815. At nine years of age he was employed as 'a trapper boy' at Penshaw Colliery. At twenty he was overman at the same place. But he had meanwhile, thanks to the precocious exhibition of an extraordinary aptitude for mathematics, held a situation in a land surveyor's office at Newcastle-on-Tyne. A twenty-four he became head viewer of Monkwearmouth Colliery, with the entire control of everything relating to the mine, which was then the deepest in the kingdom. A year later we find him part proprietor of Washington Colliery. He was soon afterwards chief mining engineer to the collieries, harbour, and railways of the late Marquis of Londonderry at Seaham. In 1863 he purchased the Penshaw Colliery, and subsequently became the chief owner of the Powell-Duffryn steam-coal collieries. To these he afterwards added others in South Wales, Staffordshire, and North Wales. He has latterly joined others in acquiring collieries in Nova Scotia, which bid fair largely to augment the world's store of fuel. But the fact of his being one of the largest colliery proprietors in England should not blind us to another characteristic of his wonderfully busy life. Not long ago, Mr. Bramwell, as President of the Institute of Mechanical Engineers, said, in an address which he delivered at Cardiff, 'Sir George Elliot, as a large employer of labour, has ever earned the confidence of his men.'

In a memoir which appeared in the *Practical Magazine*, September, 1874, the writer says: – 'It is, perhaps, not too much to say of Sir George Elliot that had it not been for his pluck and energy the Atlantic cable would not to-day have been an accomplished fact. Mr. Elliot in 1849 bought the business of Messrs. Kuper and Company, wire-rope makers, who had become bankrupt; and so great was his confidence in the concern that he offered the creditors 20s. in the pound, with interest until the money was paid, besides agreeing to pay Messrs. Kuper a handsome sum for their reversion. Mr. Glass (afterwards Sir Richard) was invited by Mr. Elliot to join him, with the view of developing the business. The firm of 'Glass and Elliot,' in 1864, was merged, along with the Gutta Percha Company, into that gigantic undertaking, the Telegraph Construction and Maintenance Company. After the failure of the first Atlantic cable, in 1856, capitalists were not to be found. Six hundred thousand pounds, to commence the renewal of the enterprise was required, and Mr. Pender, M.P., Mr. Thomas Brassey, Mr. George Elliot, Mr. Barclay, Mr. Bewley, and a few others, subscribed £285,000; but the general public kept aloof until Messrs. Glass and Elliot came forward and offered to take up £100,000 of the bonds, and to make their profit contingent on success. By a legal technicality, however, this offer could not be accepted; but on Messrs. Glass and Elliot subscribing £100,000 to a new company, so great was the public confidence in this notable firm that in fourteen days afterwards the whole sum of £600,000 was raised.' Mr. Cyrus Field, in afterwards speaking of the event, said: 'Never was greater energy infused into any enterprise. It was only on March 1, 1866, that the new company was formed, and it was registered as a company the next day; and yet, such was the vigour and despatch, that in five months from that date the cable was manufactured, shipping in the Great Eastern, stretched across the Atlantic, and was sending messages, literally as swift as lightening, from continent to continent.'

Sir George Elliot has found time for other work. In 1868 he was President of the North of England Institute of Mining Engineers. He was consulted by Lord Aberdare at every stage of the Mines Regulation Bill of 1872. The miners' advocates, Messrs. Macdonald and Halliday, have borne testimony to the hon. Baronet's warm sympathy with the working men. Mr. Halliday 'believes that if the hon. Baronet had not been absent from England the South Wales strike would have been settled long ago.' Mr. Plimsoll, in his book, 'Our Seamen' states that the fleet of coal-steamers owned by Sir George has run between the Thames and the Tyne since 1859, 'without losing a single man.'

Sir George Elliot's enterprises are matters of commercial history. He is known wherever English engineering successes are recognised; we might also say wherever a telegraph-post has been erected or a cable laid down. We find him, in conjunction with Mr. Greenfield, busy at Alexandria with a harbour and breakwater, which is to cost two millions sterling. He was the other day at Newport, opening a dock that may one day raise the port to the level of Cardiff. He is a director of a number of public companies. In the north of England – in North and South Durham, along Tyneside and the banks of the Tees, and in the West Riding – his name has a place in the hearts of the people. Sunderland and Whitby will never forget him; and there are thousands of pit-folk who will ask you defiantly 'if ye ever saw his like?' Last year, Mr. Disraeli made him a Baronet. On the 31st of last month he was entertained at a banquet in the city of Durham, when a testimonial was presented to him by the Marquis of Londonderry. The company numbered above one thousand guests; among them Lord Eslington, Viscount Castlereagh, Viscount Boyne, Lord Vane Tempest, and the son of Sir George Elliot, Mr. G.W. Elliot, M.P. for Northallerton.

The testimonial... is a magnificent dessert service in silver, manufactured by Messrs. Hunt and Roskell, of New Bond Street, from designs specifically prepared by them for this occasion. The service consists of ten pieces, most of which are grouped together on a silver plateau. The centrepiece has six arms of a rich floral pattern, with a cut-glass dish for fruit or flowers. The shaft is surrounded by three figures, typifying Prosperity, Prudence and Truth. On the base are four recumbent figures, representing Commerce, Science, Industry and Mechanics. Between the figures are panels which represent, in low relief, a view of the Penshaw Colliery and the Great Eastern Steam-Ship laying the Atlantic cable. On the others, also in relief, are Sir George's arms, crest, and motto, and the following inscription:- 'Presented to Sir George Elliot, Bart., M.P., by his fellow-Conservatives in the county of Durham, in recognition of his eminent services to the Conservative cause, 1875.' The total height of the centrepiece is about thirty inches. The four dessert-stands are similar in general style, supported by figures emblematic of arts, legislation, mining and manufactures. The plateau is of an irregular oval shape, with a richly-chased border; on each side is a medallion, supported by two reclining winged figures, the one medallion bearing the crest of Sir George Elliot with a garter containing the motto, the other medallion being engraved 'Durham, 1875.' The whole of the plateau is elaborately decorated with festoons of laurel and oak leaves, while wreaths of laurel at intervals enclose emblems corresponding with the figures on the centre piece and dessert-stands. The other pieces, which complete the service, are two corbeilles or baskets for flowers, designed in the same style as the plateau and ornamented with wreaths and festoons of laurel and oak leaves, the ends terminating in figures of cupids. These are supplemented by two candelabra, each carrying the lights, similar in style to the centrepiece, but without figures. The total cost of this very handsome service is somewhat over 2000 guineas.

It has been said that George's father, Ralph, was a shepherd in Scotland who, having lost his job, walked all the way to Durham to gain work as a miner. George was known throughout his life in the coalfield as 'Bonnie Geordie' for his countenance and good nature.

He lived in Houghton Hall, in Houghton Le Spring, County Durham, but spent the last few years of his life at 19 Portland Place, in London. He had always wanted to buy Houghton Hall, but the owner would not sell. As his granddaughter Florence Gordon later explained,

The mansion has been in the continuous tenancy-possession of the late Sir George Elliot, from the year 1844 – now almost half a century; and the lease has still some ten years to run. The old Durham family of Hutton are the owners of Houghton Hall, and their representatives always declined to sell it, so that it remains in the family, the present owner being Mr Arthur Hutton, barrister-at-law, of the Temple, London. From foundations to roof the house has been very soundly and substantially built. It is a rectangular building of three stories, and retains its mullioned windows, plain parapet, and flat leaden roof, which latter commands and extensive view of the valley of the Wear from Lambton Park to Durham Cathedral.

The foundation stone for the clock tower at the Houses of Parliament was laid on 28 September 1843. When George was an MP for North Durham, he arranged for the new tongue for Big Ben to be forged by his friend George Hopper at Hopper's Iron Foundry in the middle of Houghton-le-Spring, just off Sunderland Street. The tongue, or clapper, is inside of the bell, swinging to make it ring. The bell itself was cast by John Warner and Sons on 6 August 1856 at their premises in Stockton-on-Tees, and was taken to London by rail and sea. Big Ben is named after Sir Benjamin Hall, who was the First Commissioner of Works in 1856, and was said to be 'the prince of timekeepers; the biggest, most accurate four-facing striking and chiming clock in the world'. The tongue was removed from the bell in 1934 and withdrawn from the platform in 1954. It can now be seen in Exhibition Room 1 of the clocktower.

George Elliot was waiting at York railway station one day with a friend, on his way to London. They got talking about George Hudson, the 'Railway King'. His friend suggested they visit Hudson as he was in York prison at the time, having accrued large debts, and George agreed.

In 1827, George Hudson had received an inheritance of £30,000 from his great-uncle, equivalent to about £3 million today, and used it to become a member of the Tory Party and eventually Lord Mayor of York. He played a key role in financing the London to Edinburgh Railway and in connecting many cities in the north of England. Railways were hugely important in England at the time, and Hudson was centre stage.

With the vast riches he accumulated Hudson purchased various properties in Yorkshire, such as Newby Park from Earl de Grey, and the Londesborough estate from the Duke of Devonshire, partially to stop a rival firm building a line between York and Hull. He was elected MP for Sunderland in 1846, investing heavily in the railways and docks of that constituency. He was a generous man, hosting dances for large throngs of people, including 14,000 guests in York to celebrate the queen's birthday, as well as entertaining landowners and aristocracy in his Knightsbridge home, which is now the French Embassy.

However, his financial transactions were always rather dubious in that he tended to pay dividends to his shareholders using capital rather than profit. In 1846 alone, he proposed thirty-two parliamentary bills for railway schemes amounting to £10 million. He had a vested interest as his businesses managed over a quarter of the total lines present in England. Hudson fell from grace in 1849 and was ordered to reimburse railway companies to the tune of £750,000, paying one company £200,000 directly to avoid going to court. He was also suspected of bribing MPs, and when he spoke in Parliament on 17 May 1849 his reputation was already stained. Subsequently his share price plummeted, and the repercussions were considerable. His resignation was demanded from most of the companies where he was director, and he was forced to make up the losses. Such was the stain on his name that Madame Tussauds even took the step of destroying his wax sculpture. The street named after him in York, George Hudson Street, was rapidly changed to Railway Street, although 100 years after his death, in 1971, the name was reverted.

As an MP, Hudson was protected from imprisonment. He won a seat at Sunderland at the 1857 general election but lost it in 1859. Now deprived of parliamentary privilege at a time when living in debt was punishable by law, he fled into exile until 1865, when he was asked to stand as the Whitby candidate for the Tories. The Liberal MP, Harry Thompson, happened to be the chairman of the North Eastern Railway, to whom Hudson owed money. The people of Whitby were not fond of Thompson. He had not fulfilled his promise of linking the town with the national rail grid and had promoted Scarborough rather than Whitby as the favoured holiday destination. Hudson's success seemed inevitable.

Days before the election, Hudson was detained by the Sheriff of York and taken to prison. It is thought, but has never been proven, that this was done on the authority of Harry Thompson. Thompson lost the seat two days later to another Conservative candidate who was hurriedly appointed. Hudson stayed in York prison for three months, and it is in this period that Elliot met him. He asked Hudson what it would take to get him out of jail, and when the man gave a number – thought to be around £60,000 – George Elliot generously wrote him a cheque for the stated amount. Hudson was able to leave prison, and in return George Elliot took hold of valuable estates previously belonging to Hudson in Whitby.

Hudson was sixty-five at the time of his incarceration and many in Victorian England were appalled at the way he had been treated. His troubles were not over, either; he was out of prison but still had to live in exile as he had many more debts still to pay. George Elliot and Hugh Taylor, MP for Tynemouth and North Shields, began a subscription fund for him, giving him £600 a year. This trust allowed Hudson to have a continual source of money while being legally safe from creditors. On 1 January 1870, with the small income that he had, and with Parliament passing the Abolition of Imprisonment for Debt Act, the happy outcome was that Hudson was able to live once more in London.

On his death a year later, *The Times* was quick to remark that it was the system itself that was at fault, reporting it was

> a system without rule, without order, without even a definite morality … He had to do everything out of his own head, and among lesser problems to discover the ethics of railway speculation and management … Mr. Hudson's position was not only new to himself, but absolutely a new thing in the world altogether.

His biographer, Robert Beaumont, recalls,

He is greatly misunderstood. Whilst his financial practices were dubious, to say the least, his legacy was Britain's great rail network, which he almost single-handedly created in ten absolutely fantastic years. He had amazing energy and vision. At his height, he was the richest man in England, and *The Times* estimated that thousands upon thousands of jobs were dependant on him.

George Elliot found himself with beautiful land in the West Cliff region of the coastal town of Whitby, overlooking the sea. The West Cliff estate, with its superb ocean views, was much appreciated by George:

Mr George Elliot, M.P. for North Durham, arrived at Whitby yesterday evening in his yacht, the *Livonia*. The hon. gentleman will be entertained at a banquet, given by the inhabitants of Whitby to-morrow evening. Mr Chas. Bagnall, J.P., ex-member for Whitby, will occupy the chair, and a distinguished company of ladies and gentlemen are expected to be present.

The local people of Whitby were pleased, as an article from June 1875 makes clear:

It is satisfactory to note that there is at last some reasonable grounds for believing that something will be done, and that at not distant date, with the view of improving the town of Whitby generally, so as to make it still more attractive than it is as a summer resort ... The chief centre of the proposed operations is the West Cliff property, which has recently come into the possession of Sir George Elliot, Bart. The estate is very extensive and capable of much improvement. It has a frontage to the sea, and commands a magnificent view of the romantic coast and of the fine scenery inland... To some extent the inhabitants have, it is only fair to state, been almost helpless in the matter, because for a long term of years the West Cliff estate, the principal resort of visitors, was in the clutches of the law as a result of the tedious, vexatious, and expensive litigation between the North-Eastern Railway Company and the late Mr George Hudson. When, however, it was announced that the property had passed into the hands of Sir George Elliot, and it was the intention of the hon. Baronet to improve it, the people of the town, and especially those living on the estate, hailed the news with delight, and unhesitatingly made due approaches and voluntarily offered their assistance, and co-operation with any scheme which he might lay before them. The hon. gentleman promptly reciprocated the good wishes, and during a recent visit to the town, generously started the project by inducing the inhabitants to form a committee with the view of devising for themselves some scheme which, in their opinion, would be likely to suit the best interests of the whole town...

Many years ago there were three men living in the north of England who were fast friends. These were Mr George Hudson, Mr Ralph Ward-Jackson, and Mr George Elliot. The first made Sunderland, the second made West Hartlepool, and it would be a singular, but not less agreeable coincidence, if the last-named were to add to his other laurels by making Whitby.

George enjoyed developing Whitby. He lived at No. 8 on Royal Crescent, a street built to rival the similarly named avenue in Bath. He paid for a new church on the West Cliff, used by the multitudes of people who flocked to this charming tourist town.

The people of Whitby continued to express their delight at George's presence and sponsorship, as in this article about the Scarborough and Whitby Railway in 1875:

We understand that it is intended to ask Sir George Elliot, Bart., M.P., to accept the chairmanship of this promising company, the post being vacant by the resignation of Mr Thomas Cave, M.P. There were no doubt be a peculiar fitness and propriety in such a selection, the hon baronet having almost a worldwide business connection, and substantial influence in the commercial and monetary world. Moreover, Sir George is now closely connected with the Whitby district, and as the owner of the West Cliff Estate, which is likely soon to be further opened out and developed, he is more interested than any other man in the increased prosperity of Whitby as a watering place, and that, the completion of the Scarborough and Whitby Railway would do more than anything else to promote.

George loved living in Whitby so much that he wanted to stand as the town's MP. He built Whitby Pavilion, which opened in 1879, an attractive red-brick building set right on the edge of the cliff, with the beach accessible below. It is used to this day, boasting superb conference facilities, exhibition rooms, banqueting halls and a vast array of entertainments. It is a venue for the twice-yearly Goth festival, which harks back to Bram Stoker's *Dracula* and also celebrates the town's ruined abbey, which has been standing on the cliff edge for 1,500 years. The abbey can be reached by the 199 steps up to the cliff face from the town, immortalised in Dracula's flight in the guise of a large dog.

Bram Stoker was in fact one of George's neighbours in Royal Crescent. *The Jewel of Seven Stars*, written by Bram Stoker, was first published in 1903 by Heinemann. Set in London and Cornwall, it reflects the preoccupation Victorian England had with the East. The novel highlights the differences between, and within, the two cultures; the barbary and moral woes of the Orient, as the Victorians saw it, juxtaposed with exotic allure and splendour. There is a connection to George here: at some point during his time advising the Khedive in Egypt, he obtained an Egyptian mummy which he brought back to England. It was said that this mummy, examined by Stoker on his visits to George's house, inspired the Irishman to write *The Jewel of Seven Stars*. The mummy was supposed to have been an Egyptian princess, but later carbon dating showed it to be male.

George had married Margaret Green (born 1811), of Shiney Row, in 1836. Margaret gave birth to their first child, Margaret W. Elliot, in 1838, who was followed by Ralph in 1839. By 1842, when they were residing at Belmont Cottage in Rainton, a second daughter, Elizabeth was born, closely followed by Alice in 1843. George's second son, George William made his appearance, and his fourth daughter, Henrietta, was born at Belmont Cottage in 1848. Reported in the newspapers in 1875 is a charming anecdote about his wife, Margaret:

> Last week the Aberaman Church Sunday School children, numbering nearly 400, were generously entertained to a plentiful supply of tea and cake by Sir George Elliot, Bart. M.P., at the Aberaman Park. Table had been erected in the park, which were superintended by Mrs. Snape, the Misses Snape, Mrs. Jones, Mrs. Sims, and visitors who partook of the tea. Lady Elliot, with Mrs Pyle and children, shortly afterwards visited the grounds. Her ladyship distributed prizes to the successful competitors in running, jumping, foot balling, and skipping in the shape of balls and skipping ropes, which were highly appreciated by the recipients. After another good supply of fruit, the children marched to the lawn in front of Aberaman House. The Rev. H. Jones, curate of Aberaman, addressed the children and friends, and for the very kind reception they had received. Lady Elliot feelingly addressed the children, complimenting them on the manner they had conducted themselves, stating that this was the first time her ladyship had had the pleasure of meeting them, and expressing a hope that it would not be the last.

Another warming anecdote comes from the *Cardiff Mail* a few years later:

> Sir George Elliot, M.P., with that interest in his workmen which has always been a marked trait of his character, has recently obtained from Hyde Park, from Her Majesty's Commissioners of Works and Public Buildings, several thousand flower roots, which have been distributed amongst the employees of the Powell Duffryn Collieries. So much has the gift been appreciated, that the consignment was a very large one, the demand has exceeded the supply. It is pleasing to find that the desire of an employer to cultivate a love of flowers in the cottages of his workmen is so heartily responded to and appreciated by them.

From humble beginnings, George and his wife had been honoured to be hosted by the queen at several balls at Buckingham Palace. They also enjoyed hosting their own entertainments. 'A ball given by Sir George's son was described by a newspaper as thus:

> Mrs. Elliot's ball, at Scruton Hall, was one of the most successful social functions ever held in Yorkshire. The guests comprised not only the best-known people in the county, but many of the host's Parliamentary and other friends from a distance; and among the novelties of decoration were a magnificent trophy of flowers, specially designed for and sent from the Riviera by Sir George Elliot, and another trophy, composed of foxes' heads and brushes, gained by members of Mr. Elliot's family exclusively during his mastership of the Bedale Hunt.

Another newspaper reveals:

> One of the most cheery country balls given this year was that at Scruton on the 29th ult, when the ex-Master of the Bedale, Mr. George Elliot M.P., and Mrs. Elliot right royally

entertained their neighbours and friends. The house was filled with a mass of beautiful and sweet-scented flowers from Cannes, and the ball-room – a temporary wooden erection opening out of the conservatory – was really tastefully decorated with evergreens and bunting, while foxes' masks and brushes were placed at intervals along the wall. There were bonny faces, pretty frocks, and no forlorn damsels sitting alone upon sofas or useless men blocking up the doorways, as the first part of the evening was unselfishly devoted by the daughters of the house in looking after and assuring the enjoyment of their guests – an example it might be as well if others followed.

On the following evening the house was given over to 'Susan and Jeames', who stepped it right merrily, the house party and a few friends taking part in the revels, until supper time, when the genial master and his host's health was drunk to that good old tune 'For he's a jolly good fellow'; and I hardly think that there can have been one out of the two hundred present who did not really mean what he or she sang.

Many young fellows are declining invitations to balls and dances, voting these functions in most London houses to be insufferably stupid. The fact of the matter is, many of our hostesses are absolutely destitute of good breeding, and never trouble themselves about their guests at all. Hence their 'dances' prove flat and unprofitable, and end too soon or drag too late. Only the other night I was at a dance where my lady stood the whole time at her drawing room door, utterly unmindful as to whether her guests amused themselves or not, whilst her sons and daughters capered around the room with their own particular friends. In consequence of which inhospitable facts many young men and women never danced the whole evening; and home returning declared Lady S____'s dance to have been a distinct failure.

If I gave a dance and had a family of 'grown-ups,' I should teach them that their duty (as well as my own) was not to dance, but to see that our friends were enjoying themselves thoroughly. No young man or woman should be neglected, and by this means I verily believe my parties would be preferred by *la jeunesse dorée* to the smoking-rooms of their clubs.

Sir George was also a serious civic individual. He was a freemason, and in 1877 had been appointed by the Prince of Wales to the Provincial Grand Mastership of Freemasons of the Eastern Division of South Wales:

> On Wednesday a very large and influential gathering of Freemasons took place at Aberdare, under the banner of the St. David's Lodge, on the occasion of the installation of Bro. Sir George Elliot Bart., M.P., as Right Worshipful Provincial Grand Master for the eastern division of South Wales.
>
> Brother M'Intyre said ... when they found a gentleman whom they all knew personally, presiding over Freemasonry among those with whom he also was personally acquainted, and among whom he laboured as Sir George Elliot did amongst them, they could not but look upon him with an intensified fraternal feeling as their ruler, and hail him as such with all the warmth of feeling and loyalty at their command. (Loud cheers.) He felt it to be a very high honour to propose, as he now did, the health of Sir George Elliot, and a very great pleasure to wish him a long life and prosperity. (Cheers.) He could not desire to see Sir George more happily circumstanced than he was then – surrounded by neighbours and friends; nor could he doubt but that he would be at all times as warmly supported, because he would always be found maintaining their Masonic rights and privileges, seeking to retain their regard, and always desirous of promoting good feeling amongst them. (Cheers.) Every Mason present knew Sir George, and had long known him, and he was quite sure that, as the years rolled on, they would come to regard him as their Masonic ruler with even more affection in the future

than they had done in the past, for they would find him a man whose kindness of heart and true Masonic feeling would animate every brother of every lodge in the province in the exercise of those very qualities which he possessed in so eminent a degree. He proposed, with much pleasure 'Health and Long Life to their Right Worshipful Provincial Grand Master.' (Loud cheers.) The toast was drunk with three times three.

The Provincial Grand Master, who was warmly received, said that he spoke the feelings of his heart, when he declared how proud he felt at hearing his name mentioned so kindly, and received so warmly, as had then been the case. It was surely from no personal excellence on his part – at least he was not able himself to discover it if it were so. It seemed to him that there must be some secret worth about him which had brought this great result, and all he could say was, that he was entirely ignorant of it.

George was a serious quietly religious man. He spoke in September 1877 at the reopening of Pensher Church:

Mr Chairman, my Lord Bishop, ladies and gentlemen. I feel almost unworthy to address a meeting in the presence of the Bishop of the Diocese; and it is the first time I have had the pleasure of addressing and meeting in this county in the presence of his Lordship. I cannot, however, refrain from giving expression to my great happiness and pleasure at being present on the occasion of the re-opening of Pensher Church. This neighbourhood is to me almost classic ground, and is associated with some of my dearest and most tender, and most sacred feelings. It is here where I first learnt my catechism; it is here where my parents and many members of my family rest from their labours, and where only three years ago my dear father was buried. I am, therefore, scarcely an old man yet; and it is not remarkable that in this locality especially, and also in the county of Durham,

I should look with pleasure and interest upon the great work which the Church has done. Within my recollection I think that the Church accommodation, not merely in this immediate locality, but throughout the diocese, must have been increased at least five-fold; and it is not merely in numbers that the Church has increased, but I think I may say that in teaching, in preaching, and in general worship and real scriptural and religious feeling there has also been a very great development in the Church. (Hear, hear.) I hail this with great satisfaction. I do not myself profess to be a strong party man in the Church, because I look at the Church as a whole, in its breadth, and its scope and usefulness. It contains a variety of opinions, and that variety enables the country to have the benefit of the services of so many learned and pious men in administering the holy functions of the Church. We may take for example our Bishop and Archbishop Tate as representing one cast of thought; we may take Pusey and Keble as representing another; and then we have distinguished men such as Dr Hawley, the Dean of Westminster, and Dr Temple, the Bishop of Bristol representing a third, which I may designate as High, Low, and Broad. These all represent different shades of thought in the Church, and yet how great and grand a thing it must be that our Church is capable of receiving all these different forms of religious thought, and that there should be such combined usefulness adapted to the general tastes, opinions and convictions of the people, also at the same time that all should be consistent with the doctrines of the National Church of England. It is a great advantage in a system like that of our National Church to have a combination of such varied and general learning, piety, and a diversion of mental power, amongst the chiefs who administer the great, and important and solemn duties of our Church. It is, as it appears to me, very much like the rainbow, with its various shades, and noble and complete arch, all forming a noble whole. It is with modesty that I speak in the presence of his Lordship, who does so much useful work in the diocese – work which is recognised and admired by all who admire sincerity, conscientiousness, and hard work. But I would say that temper and forbearance, and respect for the opinions of others, ought and must pervade the Church; so that you may have the soft and hallowing influences of the good Christian feeling of all the different shades of religious opinion in the Church, so that they may be blended into that beauty and usefulness of which the rainbow is an illustration. I believe that in this combination of learning, piety, and diversity of mental power we have in our Church a strong and substantial bulwark for social, religious and domestic happiness, such as I believe the world has never seen before. (Applause.)

I hope I have said nothing which can offend or touch the feelings of his Lordship or any one present. I am aware it is a sensitive and delicate subject to touch upon. I quite understand the extreme sensitiveness which pervades the minds of many upon a subject so critical, and, therefore, if I have said one word to offend the feelings of any one, I beg that it should be considered as being unsaid, so far as relates to their individual opinion. I have no feeling myself other than that of trying, so far as my humble influence and opinion are concerned, to produce harmonious union, and co-operation in the Church as a whole, so that she may carry out to the utmost possible extent the great work that she is charged with, and that we should all co-operate together in maintaining the union of Church and State; because I believe as firmly as I do in my own existence in the union of the Church and the State as being an essential principle in our national Constitution, and also as necessary for the security of the civil and religious liberty of the people of this country, for the education of the people, and for the maintenance and continuance of Christian and religious knowledge and conduct amongst us. (Loud applause.) I therefore, hope you will join me in drinking 'The Health of the Bishop and Clergy of the diocese.' (Applause.)

The Bishop responded:

> I am much obliged to Sir George Elliot for the kind expressions
> he has used towards me. It is very true that I never before had
> the pleasure of hearing him speak at a meeting in the diocese.
> The last time I heard him speak was at the annual meeting of
> the Sons of the Clergy Society; and those around me, as well
> as myself, agreed that his speech was the most valuable of the
> many made that evening...

George funded the 130-foot tower and spire of St Mary's
Church at West Rainton in 1877 in remembrance of his late
daughter Elizabeth. He used stones that he had obtained from
the Great Pyramid of Giza for the base of this spire, as well as
for the establishment of a tomb in 1878 in the graveyard of
All Saints' Church, Penshaw, which contained the following
inscription;

> To his father, mother and brothers
> Sir George Elliot Bart MP
> In token of his reverent love and affection
> Also, to Ralph Elliot, his dearly beloved son.

Tragedy had clearly touched George's life. His first son, Ralph,
who lived at Chester Square in London, died at the tender age
of thirty-five while travelling round the Cape of Good Hope.
Elizabeth died on 29 September 1861, aged just twenty. She lived
at Houghton Hall, and died from burns when her dress went up in
flames as she prepared to go out for a party. In 1880, Margaret,
his wife, passed away. He received a letter of condolence from
Disraeli, to which he replied,

> Dear Lord Beaconsfield, Allow me to express my sincere
> thanks for the very kind letter of condolence I have had
> the honour to receive of your Lordship – it has been a
> great consolation to me and my family treasure it highly:

What can I say or do but bow to the will of that great being alone. Wisdom and goodness, I dare not venture to contest.

After Margaret's death, his religious endowments continued. He commissioned a window in Durham Cathedral, to the east of the main north door, depicting three scenes of Bernard Gilpin, the Oxford theologian who lived between 1517 and 1583. Gilpin was a leading figure in the emerging Church of England, Known as 'the Apostle of the North', he was appointed rector at Houghton-Le-Spring in 1568. Particularly poignant is the bottom scene depicting Bernard Gilpin founding a grammar school at his parish in Houghton-le-Spring, which he did at a cost of £500, allowing the poorest children a chance to gain an education. So popular was the institution that he had to renovate part of his own home for use as a boarding house. Promising pupils were also helped through university. It is very touching that George chose this scene for the window, as it shows his gratitude in the life he led. It was produced by Clayton and Bell, one of the greatest Victorian stained-glass workshops. There is a horizontally set brass plaque below the window, which is inscribed, 'Sir George Elliot Bart. gave this window in 1881'.

In 1882 George bought some land in Aberaman and set about building a church there in commemoration of Margaret and their daughter Elizabeth. St Margaret's Church was finished in 1883. As well as the church he built in Whitby, he further bestowed the stained-glass window showing the Baptism, Resurrection and Ascension at All Saint's Church in Penshaw in 1889, in remembrance of his brothers and son. He provided the lychgate archway to the Houghton Hillside Cemetery as well as the dramatic Great East stained-glass window at Houghton Church.

George did not remarry after Margaret's death. He was, however, involved in a well-publicised case ten years later, when the professional singer Emily Mary Hairs sued him for breach of promise. She claimed that he had promised to marry her, and she wanted £5,000 in damages. The case was rejected by the court.

A few years on, in 1886, George built the Elliot Home for Seamen in Temple Street, Monmouthshire, where he was MP from 1886 to 1892. This large stone building was an establishment and a church for sailors, with a resident vicar, run by the Mission to Seafarers Society.

After a long and eventful life, George died at 3.10 p.m. on 23 December 1893, aged seventy-nine, suffering from acute pneumonia. He was buried at the Houghton Hillside Cemetery, with the funeral being held on 28 December 1893. The family vault is also the resting place of his daughter Elizabeth and his son George, along with his grandson George.

His second son, Sir George William Elliot, died on 15 November 1895 aged fifty-seven. He resided at Scruton Hall, Northallerton. George William's son, the third Sir George, lived in Hanover Square, London. He was born on 30 May 1867 and died on 14 October 1904. He was closely followed by his younger brother, Charles, the fourth baronet, who died without heir in 1911.

The vault at Houghton-le-Spring was built in 1862 after a Church of England faculty was issued. Originally it had an outer door, which opened into a large inner room with shelving and a beautiful tiled floor. Sadly, it was vandalised in 1957 and had to be resealed.

When George passed away, he had been working on a plan to bring together all the coalfields in the country. He wanted to better the working conditions of the miners by collecting a certain percentage of the industry's profits into a retirement fund for the colliers.

In giving tribute to him, his granddaughter Florence Gordon explains:

Exactly a month ago, Sir George was in Newcastle, when he attended a meeting of coal owners in connection with his proposed Coal Trust scheme. He was then in excellent health and spirits. Some few days afterwards he attended Lord Salisbury's meeting at Newport, and, after leaving a much over-heated room, had to wait a good while for his carriage. He caught cold, and gradually grew worse, until, for the past three or four days, his condition was evidently hopeless. His only son, George William, is member for the Richmond division of Yorkshire.

The late baronet's personal characteristics were as noteworthy as those which brought him commercial success. No better epitaph could be written of him than in the words of the motto of his house, '*Labor et veritas*' [Labour and Truth]. Sir George Elliot was energetic in his dealings, and his success in life was due probably as much to his honesty of purpose as to his never wearying energy. Nor did he spurn the ladder by which he rose to fame. On the contrary, he was ever willing to stoop down and lend a hand to men who were struggling up the way which he had ascended. His very presence breathed kindliness and encouragement, and his beaming countenance attracted irresistibly all people with whom he came in contact. In our own neighbourhood the death of Sir George Elliot removes an important link between the past history of the coal trade and the present – a history which he, together with the late Mr. Nicholas Wood and others, did much to build up and perpetuate. His recent efforts in the direction of forming a great coal trust are still fresh in the memories of most people. Sir George Elliot was a captain of industry in the true sense – a man to whom labour owes more than it is sometimes inclined to repay. His was a life, too, with its lessons for everybody: his, an example which anybody might seek to emulate with advantage to himself and to the State.

Sir George was remembered in *Vanity Fair* as 'a very keen man of business ... simple, unaffected, and cordial, yet without pretension ... his manners good, and his companionship as pleasant as for those who desire to inform themselves it is profitable … when he [was] in England few men [were] more congenially welcomed by those who [knew] him'. *The Times* remembered him as well:

> The public loses in him a man who had the capacity to place himself at the head of important moments and the energy and industry to continue to direct them with success during a lifetime which has lasted for nearly 80 years.

INDEX

Queen Victoria 21, 82, 118, 147,
    181, 183, 207, 208, 210,
    211, 212, 213, 223, 263
Queenstown, Ireland 131, 132,
    133, 134

railways
    high-pressure steam engine 54
    puddled iron 54
    rolling mill 54
Redcliffe, Lord Stratford de 177,
    180, 182
Rhondda Valley 83
Rhymney Railway Bill 77, 91
Riska colliery 45
Roman Empire 166
Rothschild, Baron
    James 182–183, 211
Rothschild, Baron Lionel 209,
    211, 213
Royal Commissions on Coal
    Supply and on Accidents in
    Mines 70

Said Pasha 171, 174, 175, 176,
    177, 180, 181, 185, 186,
    187, 188, 190, 191
Saint-Simonians 180
Savery, Thomas 53
Scott, William 38–39
Second Industrial Revolution 50
Shiney Row 10, 12, 67
Siemens, Dr Werner von 110
Slater, Samuel 64
slavery 37, 61, 192
Smeaton, John 53
Smith, W. H. 220
Smith, Willoughby 111, 113, 149
Snow, John 56

Society for the Study of the Suez
    Canal 172
Society of Arts 108
Sopwith, Thomas 68
South Wales 71, 72
spinning jenny 50, 51, 54
SS *Elba* 114
SS *Great Eastern* 156, 157, 158,
    159, 161
St Margaret's Church,
    Aberaman 276
St Mary's Church, West
    Rainton 275
Stanley, Edward, Henry 210
Stanton, General 210, 211, 212,
    213, 215, 243
steam engines 52
Stephenson, George 68
Stephenson, Robert 68, 172, 180
Stoker, Bram 267
Suez Canal 149–50, 164–222
    Suez Canal Company 180, 183,
        185, 194, 195, 196, 198,
        199, 201, 202, 207, 208,
        209, 214, 220, 252
    launch party 199, 201
Sweet Water Canal 188, 190,
    194, 195
swing riots 52, 60

Taff Vale Railway 73, 83
Technical Advisory Committee on
    Coal Mining 70
Telcon (Telegraph Construction
    and Maintenance
    Company) 150–151, 159,
    258
Tewfik, Khedive of Egypt 252
Thames and Severn Canal 54
Third Republic (France) 208